ONE FOOT AWAY FROM A MILLION

BY

DONALD E. MCCOY

McCoy, Donald E.
One foot away from a million / by Donald E. McCoy. -- 1st ed.
p. cm.
ISBN 0-9665584-6-4

1. Gold mines and mining--History--20th century--Fiction. 2. Airplanes--Fiction. 3. Mexico--Fiction. 4. Los Angeles (Calif.)--Fiction. 5. Colorado--Fiction. 6. Atlanta (Ga.)--Fiction. 7. Historical fiction. I. Title.

PS3613.C38385O64 2003 813'.6
QBI02- 701389

Cover by J acquie Yost

Bluwaters Press
P. O. Box 878
M esilla, NM 88046

Printed in the United States of America

10 9 8 7 6 5 4 3 2 1

Dedicated to the small miners and
mill operators who were destroyed
by government regulations.

Cover photograph
by Donald E. McCoy

The Golden King Tunnel near Jacksonville, Oregon, is part of the Oregon Belle Mine that was developed in 1880. The Golden King Tunnel connected to the main quartz vein and was completed about 1888. Records reported that by 1905, $250,000 in gold had been taken from this mine. The mining was done with steam drills.

Introduction

I recall, back in the 'good old days', belonging to the boy scouts. A friend and I borrowed a tent and backpacks and walked into the mountains for a four day hiking trip. This was supposed to give us a badge. For several hours, we struggled up steep trails. In spite of the heat and threatened dehydration, we forged on, looking for the perfect camping spot. Finally, we stumbled upon a beautiful little glade with trees, ferns and a small stream burbling by. We had a great time fishing, swimming and cooking over a campfire. My buddy woke me one night and said there was a big mountain lion in the tree glaring at us. This was a great adventure for two kids getting so far away from civilization.

Two weeks later we drove up one of the many mountain roads in the area. We found a walking trail leading into the hills. Walking about two hundred yards we discovered our camping spot that we had thought was so far away from the world. In many ways this has been the way my whole life has unfolded.

I was brought up in a family that had many stories and legends to pass down. My grandfather had been a prospector, an uncle had joined in the Alaskan gold rush, and another uncle had been a captain in the Arizona Rangers, just to name a few. From an early age, I felt this was the way things were supposed to be: forging through the jungle, struggling up a snow covered mountain, always with a wild Indian lurking in the background, or a rattlesnake in your bedroll. There was never any mention of how much money these pioneers had hoped to make. The bottom line was the adventure, seeing and learning new things. It has always been more interesting to wield a machete and chop a path through a jungle than to take the freeway to the get to the same place.

This story is about chopping through jungles, climbing mountains just to see the other side, and running into constant modern day obstacles that were different from that which my ancestors had to contend. I spent many years in Mexico at a time when the gold peso sold at their banks for $62.50. Had I taken the money spent on interesting adventures, and bought the coins instead, I would be quite rich today.

Looking back, I am very glad I opted for the adventures.

Donald E. McCoy

CONTENTS

PROLOGUE

This story is about people we have all known. The story relates to the exuberant pupil in the classroom who manages to get sent to the corner, on a stool, most every day he appears in class with the teacher's warning, "You are not paying attention, Johnny!"

There are varying degrees relating to Johnny's sentencing. Some days he sits on a stool watching the class, other days he faces the wall. If it is really a goof he gets a pointed hat called a duncecap. The worst daily offense caused by his errant ways—from the teacher's viewpoint—would be the stool, the wall, and the cap, plus a big sign around his neck that spells out DUMMY.

To expedite the sequence, when the teacher is very busy, she or he only has to say, "Johnny, here's your SIGN".

MEXICO

The Sonoran Desert is a beautiful garden of weird cactus, rocks, sand, and the occasional sidewinder or gila monster. The colors and shadows dance in the heat waves. Once in a while a mirage will seem so real you believe it possible to take a boat and go fishing in the bright blue lake it projects onto the desert floor. There are few roads and fewer people, most of whom are Serie Indians who still speak their own distinct language despite the Mexican-Indian border lingo.

I was sitting on a rock looking out across the Altar desert toward the Sea of Cortez. There had been abundant rain this spring. It was a rare picture of a veritable garden of beautiful plants and flowers. It is amazing how seeds can lie dormant for so long and still be able to spring back to life with the addition of a little rainwater. Soon this picture will return to shivering heat waves and blowing sand. Better appreciate this, it only happens once in twenty years or so.

"Why are you doing this Eugene MacDonald?" I was talking to myself. When you live alone long enough, you tend to talk to yourself. If you are paying attention, you even find your own voice answering. "Well, you wanted to be in the gold mining business, didn't you? Sure, but it does get lonesome." After you get through the initial excitement of beginning a new project, the romance seems to disappear.

I had been working here at the Sierra Pinta Mine for about six months with no one to talk with but the Mexican workers. They did not speak much English, so I felt justified in talking to myself.

The mine originally had been developed by an English company in the late 1800's. The English had located and developed the whole layout including a narrow gauge railroad that ran from the coast of the Sea of Cortez up to the mine. The English had backed out when Pancho Villa was raising hell along the northern border, during the upheaval of Mexican politics around 1914. The local history surrounding the mining company was that the owners had given title to the mine to their Mexican attorney to hold until the politics settled down. When the wars were over, the attorney refused to give the mine back. A rather crooked move on his part, but the dream of gold does that to some people.

There had been a long history of promotions since. The old timers must have overcome stupendous problems building and supplying the mines with only wagons, mules and in the case of the Pinta, ships. They had laid the tracks on a bed of tailings dug from the mine. The roadbed at the time contained more gold than most mines have today. The concentrating plant was located on the coast, and they shipped the resulting material to Wales for final processing. This was an indication that the ore must have been very rich to go to all that trouble. Wales was on the other side of the world.

The main shaft had been mined on an incline of about forty-five degrees. There were no timbers as the rock was very hard. In mining lingo the rock would be termed competent. The shaft section was close to five hundred feet deep and at the bottom was a sump full of water. I had gambled that this water was primary water and had not accumulated from the surface.

Manuel and I had installed an electric pump in the sump and piped the water to a holding cistern on the surface. A diesel generator supplied the power for the pump and for the lights in the tunnel.

About three hundred feet down the shaft was a cross cut drift that went through the mountain approximately fourteen hundred feet to the outside. This tunnel followed the quartz vein of gold and was mined above and below on the dip of the vein. The tunnel through the mountain produced a constant draft as the wind from the ocean blew through the opening. The mine acted as an air conditioner, making this the coolest spot in the desert.

The miners had used ocotillo cactus to shore up the waste rock that was stacked on the incline above. After so many years, this was becoming dangerous. If the ocotillo barricade failed, several tons of rock would cascade onto your head. Every few feet there was a shrine to honor a miner who had been killed at that spot.

There were funny animals of the raccoon family called coati mundi. They would steal eggs from the kitchen and hide them in the mine. I always wondered how they were able to carry the eggs that far without breaking them. They seemed to do it as a joke. One time I found a loaf of bread still in its wrapper tucked behind the loose rock.

Manuel was coming up the hill, picking his way between the sharp cactus. He had married a lady of the evening and brought her with him when he took the job helping to rehabilitate this old mine. His wife, Sofia, was the camp cook. She had a gold front tooth that added character to her smile. The tooth was a kind of bellwether. When I did not notice the tooth anymore, it was time to go to town.

I had arrived late last night with another load of supplies and had slept later than usual. "Buenos dias, Senor MacDonald," Manuel called from below, slipping on the rocks. "Did you get my message?" He asked in his border dialect. It was not Spanish, but close enough that I could understand him most of the time.

"I saw a note on the screen. But I didn't read it. It was too dark and I did not light the Coleman lantern. What was the message, Manuel?"

"Yesterday afternoon I saw a big rattlesnake going under the door and into your house. I put the note there to warn you!"

Shit! Where is the romance in this? I had slept all night with a damn snake. Well, I guess if the snake did not care, it did not bother me. Up to now my main form of entertainment had been keeping cadence with the tick-tock of Greenwich Mean Time that I was able to get on short-wave radio in the evenings. Snakes would add to the excitement. "Thanks for the warning, Manuel."

The trip to Tucson had taken four days with Merlin. I had decided to name the truck Merlin because it was able to do magic things, such as take me through the cactus and not get a flat. I had delivered samples to the lab for analysis and returned with a propane refrigerator.

The officials at the border had tried to play their pay-off game. I was usually able to talk my way out of giving them any money. Once you give those guys the first peso, you pay more each time you cross. I had arrived after the customs office had closed and the officials wanted me to wait until the next day. Then they would charge me an exorbitant fee. This fee would depend on how much tequila they had the night before.

I parked the truck, sat on the curb alongside of them, and watched as they collected cash from the people to allow them to cross the border. Actually this was the way they were paid. The Mexican government did not have much going for their payroll. Soon they got tired of my watching and suggested they would not notice if I wanted to get in my vehicle and go away.

"What's been going on, Manuel? Get anything done underground?" I asked.

"I haven't worked in the mine, Senor. I don't like to go down there alone."

"Where are the other workers?"

"They got a ride back to Sonoita with Senor Gillespie. They said it is too hot to work out here," Manuel replied. On October 31st it had been 130 degrees Fahrenheit.

"Well, that will cut some overhead, we're about done cleaning up anyway."

"I've been cleaning out the icehouse, thought it would be a good place for storage."

The original mining company had built an icehouse where they made block ice. The double wooden walls had been filled with sawdust for insulation against the extreme hot temperatures. Even today, ice was a luxury hard to come by in this part of the world.

Manuel was a hard worker. He could always find something to do, but he would not go into the mine without me. He was more afraid of leaving me alone with Sofia than he was afraid of the dark.

"There were some officials from the federal government here on Thursday, Senor. They asked a lot of questions about you."

"Oh? What did you tell them?"

"As little as possible. I did not like them. They weren't the local officials I know. I told 'em that I didn't know much, only that you were a mining engineer from Arizona. They asked about dynamite, too. I did not show them where we hid it." I did not like the sound of this. Anytime you get a group of officials nosing around unannounced the chances are, there is trouble afoot. "They said they would return another day. I think they wanted to check your papers."

"Uh huh. Have you seen my partner, Roberto?"

"No. He hasn't been here at all. Sofia has made fresh tortillas. Come down for breakfast, Senor." Montgomery is a big help, I thought, why have a partner, if he does not do any work. I had learned about the mine from a friend who knew Robert Montgomery. My friend had told me what an honest fellow Robert was and he was supposed to own the right to the mine. Montgomery was originally an American citizen and had married a Mexican woman. This gave him the right to work in Mexico. He had brought a lot of machinery and tools across the border for me. Some of the items, like mercury and dynamite, were easier for him to get across legally.

After a good breakfast of huevos, frijoles refritos, chorizo and tortillas with lots of salsa and strong boiled coffee, Manuel and I unloaded the refrigerator. A gas refrigerator in the desert is a luxury beyond imagination.

"Manuel, I am a little worried about the officials that were here. I think I should go back to Tucson and try to find out if they were really official. I can probably get my friend Wilbur to call the governor's office in Hermosillo, to see if they know if anything is going on that we do not know about."

By the time we unloaded everything, set up the gas bottle and hooked it to the fridge, it was late in the day, but I headed north anyway. When I crossed the border, the guards looked at me a little strangely. They were used to me spending three to six weeks at the mine.

Driving in the desert alone gives you a chance to think. In this part of Arizona there is not much traffic even in the daytime. I was reminded of the of old adage handed down from the gold miners of the past. In mining you finish a day drilling, mucking, crawling and cussing, your muscles all ache and you can't stand straight. You feel as though you are a million feet away from a dollar. The next morning, watching the sunrise and listening to the birds singing, you know without a doubt, you are only one foot away from a million. I had never believed I had gold fever, but I could relate to the story.

I got as far as Why, Arizona, and decided to call ahead and see if Wilbur would still be at the lab. Wilbur Wright owned a complete assaying laboratory. He could analyze any inorganic material. He had an atomic absorption testing facility, called AA, as well as the fire assay procedures. Most of the work he did was for the big copper mines in the area. Wilbur was a certified chemist and well respected in the mining community for the accuracy of his results. I had used his facilities for some time checking samples from various mines in the southern Arizona area. We had become friends. Wilbur had even hired my son to work in the lab.

Wilbur answered after about eight rings, "Wilbur, how late will you be working tonight?"

"Should be a while yet, where are you now?

"Why."

"Cause I want to know, dumb shit," and hung up the phone.

I had to get the operator again and called him back. "Why did you hang up, Wilbur?"

"Dammit, you were being smart with me and asking why I wanted to know where you were."

"I said I was in Why. Maybe I should ask them to change the name. This is the place where the mailbox is on a long pole and the guy calls it airmail."

"Oh yeah, now I know what you are talking about. You are just south of Ajo. I'll be here. See you in about an hour and a half."

I found myself thinking of what it meant to have gold fever. Over the years I had heard stories of weird things happening to people who were exposed to the metal for long periods of time. The movie, Treasure of Sierra Madre, with Humphry Bogart playing a man who went crazy was a good example of this influence. In the beginning of the story, he swore he would never be affected, but he was. Since the body is made up of electrical ions, I suspected that the problem was caused by the frequency of the positive ions given off by the metal. This frequency affects people in different ways.

A lot of studies have been done on people and exposure to electrical ions. The studies suggest it could have been the reason the pioneers on the prairies went crazy, especially the women. The winds from the mountains carried an abundance of these positive charged ions and the women in the houses were apparently subjected to a high concentration. In some instances it was so bad, the women would murder all the kids. Psychologists have shown that if a person has the opportunity of possessing a bar of gold or a bar of platinum, which is worth far more money, they will take the gold. Pretty interesting stuff. Glad I do not have gold fever.

When I arrived at the laboratory, Wilbur was just finishing up the washing of the beakers. "Glad you came back. Let's go get a drink I have something to tell you."

"Okay, but first do you think we could call the governor's office tomorrow? I want to find out if the officials who showed up at the mine this week are legitimate."

"Yes, I just heard today the gov is in town. Let's go. Get your coat."

A shot and a beer was what I needed after driving that far twice in two days. "What did you have to tell me, Will?"

"A couple of customers were in yesterday looking for help on a mine they have south of Chihuahua. If you want to go, they will be leaving day after tomorrow. Their names are Clay Coughenour and Dick Nichols. Dick has an airplane and a pilot already lined up. Also, I am sending a highly rated geologist from Canada to take the ore samples."

"That sounds like an adventure. If I go along, could you try to run down what kind of problem may be brewing at the Sierra Pinta?"

"Sure, the governor is a friend of mine. If someone is trying to pull a fast one, he will tell me what is going on, if anything."

"Thanks for the beer Wilbur. I'd better get going, I'm kinda tired. Give the Clay guy my phone number. We'll coordinate from his place."

We were flying at seventy five hundred feet, over the Sonora Desert, on the way to look at the gold mine in the Sierra Madre Mountains, Mexico. Shortly before, we had cleared customs at Nogales. There had been some questions on the ownership papers on the plane. They had been forgotten when fifty US dollars had changed hands.

This was a nice looking aircraft, a TU206 Cessna. The 'T' for Turbo and the 'U' for Utility, which accounted for the double doors on the right side of the air-frame. The wing span was forty feet. The top of the fuselage was red and the wings and underbelly natural aluminum. The three hundred horsepower engine had just been rebuilt and a 310-nose gear had been installed for extra strength in landing on rough terrain.

The six cylinder Continental engine pulling a double bladed constant speed propeller, emitted a throaty growl and, with the turbocharger, could fly up to thirty thousand feet. If you did not leave the oxygen bottle in the hanger, the pilot and passengers could too. Later, when we were to know him better, this flying machine would be christened MORTIMER!

I was in the rear seat next to Clay. The geologist was in the middle seat. Harvey and Dick were the pilot and co-pilot. I was to learn later, Dick had no license. The reason the plane had a new nose gear was, about a year ago, Dick had crash-landed it somewhere in Baja, California, trying a forced landing because it had run out of fuel.

Dick could fly because he owned the plane and besides, who needs a pilot's license in Mexico? Harvey, a licensed instructor, was supposed to be doing a little overseeing on Dick's ability to drive an aircraft. However, he had been out very late the night before and was now sound asleep.

Our destination was somewhere in the mountains, on the Chihauhau-Sonora border, about two hundred miles north of where Bogart found his treasure in the Sierra Madre. The trip was to take three hours from Nogales. The 206 Cessna has a range of four hours at this altitude, assuming a full load of fuel of eighty gallons.

There were two radios, but of no use here in this part of the world. An Automatic Directional Finder (ADF) was the only navigational guide, other than the gyrocompass, and a hand wound clock. The ADF was useless, as there are no radio stations to key to.

The flying done in this area is by the seat of the pants. In other words, you are supposed to know where you are going. Dick had flown down here once and I thought he knew the way.

While Dick was 'knowing' his way, Clay had a chance to give me some background on some of the local history. Clay's brother, George owned the cattle ranch where the mine was located. In the days before transistor radios, Clay's family would buy gold from the local miners at a low price. The miners, unaware of prices of gold in other locations, would sell at the locally prevailing price that had not changed for years. When the miners acquired a radio and learned what the competitive gold prices really were, that business went away.

About two hours had elapsed since departure when I thought to ask a simple question, "Clay, you have flown down here a lot, correct?"

"Sure, I've ridden mules and had to walk most of the time."

"Okay then, if the ranch is supposed to be in the mountains, why after all this time don't we see any mountains?"

"Oh shit! What's Dick been doing? Hey, Dick what's happening up there? Are you jacking off or reading Fate is the Hunter?"

Dick yelled back, "Harvey is asleep, I have been quiet so as not to disturb him."

"Give me the charts, asshole," Clay growled. After a couple of minutes, Clay said, "It looks to me like we're about one hundred miles too far South."

"So there's no problem, right? We have fuel for another couple of hours?" I had to ask.

Clay snorted, "With this payload, we didn't fill the tanks. We don't have enough fuel for any mistakes. There's gasoline at the ranch. We had planned to fill up there for the return trip."

By now Harvey is among the living and beginning to realize he has a much larger headache than before he started his nap. He snapped, "Dick, why are we in this location?"

"I've been looking for the needle shaped peak that Clay told me to aim for, but have not been able to spot it."

"Give me the controls Dick and start looking for a smooth spot. We're very low on fuel. We've got to get this bird on the ground."

Harvey started a wide 360-degree bank. Now there are six people looking for a soft spot to attempt a controlled crash. Actually, when you think about it, any landing with an aircraft is essentially a controlled crash.

After about ten minutes of circling, I caught a flash of sunlight from something. Could have been a window off to the northeast, about five miles. Sure enough, there it was again. Flashes of sunlight like a signal beacon. Harvey caught sight of the second flash. He brought us down to about 500 feet. We could see a lot of sagebrush, big rocks and an old wooden building with a corrugated roof. There appeared to be a cleared area that looked useful for a landing strip.

Over the noise of the engine, Harvey called back to Clay, "Do you know anything about this place?"

"I'm sure I've heard my brother talk about it. The ground should be solid enough to land on."

Under these conditions, it takes a good bush pilot to judge all of the hazards presented. There is no way to determine the wind direction or if the terrain will be too soft to support the weight of the plane. Harvey banked, lined up and began his approach on what we hoped would prove to be a runway.

It is very quiet. The dust had settled and I was sitting on a bale of straw near the building thinking about the Little Johnny story from the past. I was wondering who should get the Dummy Sign, maybe Dick? Maybe Harvey?

Inside the building, with the bales of straw, was another airplane. The sun bouncing off the windscreen of the plane had flashed the beacon. Any earlier or later in the day, the sun would have been in the wrong place to reflect the signal to us.

"Clay," someone asked, "what is this plane doing way out here?"

"You don't want to know. When we get out of here, you better just forget about it."

Behind the straw covered with an old tarp were some barrels. Harvey kicked one and found it was full. Dick retrieved a wrench from the 206 toolbox and they opened the drum. Harvey dipped his fingers and smelled the liquid. He turned and said, "Well, looks like we have located a gas station."

In frontier areas like New Mexico, Arizona, Texas, and especially Mexico, people make do with what they have. One does not call Ace Hardware and order a funnel or pump to be delivered. I had seen tires stuffed with rags and cactus used in place of a broken spring. One time I saw a gas tank mounted on top of the cab of a truck. The fuel pump had failed and gravity now fed the gas to the carburetor.

They found an old bucket covered with sand at the base of an ocotillo cactus. In this part of the world it takes a long time for metal to rust. There is very little moisture, and if it does try to rain, the rain tends to evaporate before the drops can hit the ground. For this reason the bucket did not have any holes in it.

The boys fashioned a funnel from the cardboard box that the groceries were in. Harvey climbed over the cowling and on to the wing. The rock expert helped tip the barrel. Dick caught the spilled fuel in the bucket and handed it to Harvey. Harvey poured it in the cardboard funnel about two quarts at a time. This reminded me of the story of the ant moving the pile of wheat!

Some time back, I had flown on an Aeronaves de Mexico, DC6. In Tuxla Guttierez, in southern Mexico, they had refueled the plane with hand pumps from barrels on a donkey cart. But this was slower. We were in Mexico, thus, everything was correct. Mexico had invented 'slow' a long time ago.

While the ant was moving each kernel of wheat, Clay and I had some quality time together. Here was a guy that you would never notice in a crowd. He was balding and had a splotchy face and a red nose, as though he drank too much. He appeared to be in his late fifties. His movements were slow and deliberate, giving a feeling of 'I don't give a shit'. He had said that he was working as a maintenance supervisor for a mining company near Tucson.

"Clay, does Dick have any mining background?" I inquired.

"No. He's a partner because he owns the plane. He's having trouble with it though. After the accident in Baja, his wife wants him to sell it. It cost a bunch to dig it out of the sand dune."

"In order to work down here you really need a plane, don't you?"

"Sure do! This one is what you want. It will carry its own weight and more if you don't follow the factory recommendations."

Becoming more interested in the plane, I asked about the engine time.

"Oh, no problem. In fact this is the first trip since Dick rebuilt it. He discovered a cracked crankshaft after the prop hit the sand dune. These Continental engines are rated for 6500 hours, but with Dick being the mechanic, who knows."

The desert is a beautiful place. But mere man cannot fight Mother Nature, you have to learn to live with her. You must treat her like a nice lady with PMS. You never really know what is coming next. It was late fall and today was most pleasant. The air was cool and clear. Once in a while a little dust devil would skitter around, then quickly disappear. This was in our favor. If it had been very hot, we would have much difficulty getting airborne again especially with our load of people, tools and food.

I said to Clay, "Your brother has lived here a long time. What are his flying rules?"

"One cardinal rule, when you're in the mountains, fly only before ten in the morning. Some times on days like today, you have a chance just before dark. Usually these are the best times for low turbulence. Another rule is never go after dark, particularly with a single engine." The ants had transferred about five gallons of gas. We should take on a minimum of thirty.

"How far is the ranch?"

"Good thought," said Clay. "Hand me the chart, would you?" He scaled from our present location and was able to calculate about an hour and a half.

"It will be late by the time we get enough fuel on board. We will be in danger of running out of daylight."

"What are the conditions there?"

"Well, you can only approach from the east. If we miss the approach, you can only turn north. If you turn to the south it is a box canyon and there is no way out. In fact, just last month there was a guy by the name of MacDonald who turned south and crashed up there. Killed him dead. The strip is long enough but there is a creek running across it. The fun part here is to find the bridge that crosses the water. The old wooden bridge is hard to see in the high grass and is barely wide enough for the landing gear. If they haven't fixed the fence there will be a cow or two eating the grass, hopefully not in front of the bridge. Also, if George is home his plane will be in the way," Clay added.

"Will Dick be able to land in there?"

"No sir, not Dick. Harvey can do it. He's okay."

"Well," I suggested, "tell him about the bridge."

This was really a pretty place. The mountain was surrounding us. High up on a purple peak were the lingering shadows from a disappearing sun. There were no cows on the field. Harvey had made the correct approach and the bridge, though narrow, had been crossed.

From the east we could see another plane lining up on final approach.

"That's my brother", said Clay. "Wonder where he's been flying off to?" Soon another Cessna, a 182, rumbled across the bridge, spun around and coughed to silence. The first thing out the plane door was a pair of cowboy boots attached to a very tall, gray haired fellow, who was quite drunk.

"Hola amigos. Was short of tequila. Decided to run down to the liquor store to get you some more." With that said, he fell down in the grass.

"Meet George," Clay said.

Now, I knew George got the Sign. Maybe instead of the Stetson hat he should have been wearing the dunce cap! Another rule is never fly with a pilot that you don't know.

After a filling dinner of tortillas, frijoles and fresh venison, I found myself sitting near George in front of a large fireplace. Some of the surrounding rock had small seams of gold smiling and winking from the light of the dancing flame. George had his lanky frame stretched out and actually looked like a large great Dane. He say, "Mac, how do you come to be in Mexico?"

"Well, I got into a partnership with a fellow near Rocky Point, at the old Sierra Pinta Mine."

"I know that area, an American name of Gillespie had some development there." George replied.

"I've met Gillespie, he was trying to sell me an old Chilean mill. It was too large for what we're trying to do."

"What's your partner's name, Mac?" George asked.

"Robert Montgomery. I'm having a little trouble with him right now. I'm not sure he is a Mexican citizen, so I'm having an attorney in Mexico City check him out." At that time, when you worked in Mexico, their business laws required any foreigner to have at least a 51% partnership with a Mexican citizen. The 51% belonged to the Mexican of course. "How long have you been down here, George?"

"Long time, Mac. Our father was a mining engineer and I was born in Mexico. The other kids were hatched in the States."

Clay handed me another glass of tequila. It was going down pretty good after the misadventures of the day. "George, Mac is interested in buying Dick's 206. Why not tell him how you learned about planes?"

George frowned with his wrinkled face and said, "Back in the twenties this country was a world of its own. There was no communication, only wagon roads, burro and mule trails. Our old man figured that he would have an advantage if he could get around more quickly. The family was all here then. Since I was the oldest they chose me to go to El Paso and take flying lessons. The old man told me when I felt I could fly pretty good to buy a plane and bring it back.

"Down here the Indians had never seen a plane, if they even knew what one was. So, I said, 'Good idea dad, but where do we land an airplane?' Dad grumbled, 'don't worry about that. When you get back there will be a place to land'."

Mexico was the original land of manana, nothing ever seemed

to get completed. Yet, here he was traveling at least 400 miles north to buy an airplane, which in those days was essentially a combination of wood and canvas, learn to operate it, then find his way back into a mountain wilderness. He could not know if there would be a safe place to set it down.

"Christ, George, you must have had some faith in your dad."

"Even though he was a miner, I had to trust him. He was El Jefe and besides my mother liked me. She wouldn't let him forget to build a strip."

"So, Clay and the Indians cleared a strip while you took a ride to El Paso. How long were you gone?"

"Took me about three months to get the hang of the thing. It took two days to fly down here."

"How did you know where to go? Things look different from the sky."

"They could hear me quite a way off, so they built a big fire and I followed the smoke." That seemed simple enough.

"Mac," George advised, "if you're going to fly down here in these mountains, always go with someone who has been in and out of a particular field several times. Do not attempt it on your own the first time. The best bet is to go with one of the local bush pilots. Watch what he does. Then go get your machine. Flying in these windy conditions you're going to pucker the seat more than once a day."

I was to learn later from his obituary, that as a young man he had been conscripted into the Mexican military and had the job as President Mateo's personal pilot, flying DC3's.

The big oaks and cottonwoods were stretching their limbs and the slight morning breeze was drying the dew off the leaves. The only sound was in the distance coming from a small generator making electricity for somebody. On the other side of the pasture I saw what appeared to be a mill building. Crawling through the barbed wire, I made my way across the field to the building. Inside I found Clay and Dick working on a little jaw crusher. "Buenos dias, amigos."

"Hi Mac. How'd you sleep?"

"Super, not one train came by to wake me up."

The machinery was antique, a jaw crusher, ball mill and a bank of flotation cells, all in poor repair. It looked as though it had not been in operation for a long time. I asked Clay, "What is the capacity of this mill?"

"About 20 tons a day with the mineralized quartz." Then added under his breath, "If we don't bust something."

"Why aren't you operating?"

"We need to increase the volume of concentrate. The smelter will only give us a contract if we can guarantee them a minimum of five tons a month." The smelter was near Ciudad Chihuahua, about 200 miles to the northeast. "If we can get the financing together then we can upgrade our volume and get an advance payment on the shipments." Clay added.

"What is the value of the concentrate?" I asked.

"The ore is averaging about 15 grams a ton and if we maintain a twenty to one ratio, we will get about eight thousand dollars a month on contract," Clay calculated.

The official government price for gold in the United States had recently increased to about forty-three dollars per ounce. Here in Mexico the price was two thousand five hundred pesos per kilo or one hundred US dollars per ounce. Considering the low wage scale this looked liked a good business proposition.

"What would be the cost to increase the volume of the mill?"

"We should get into operation for about $10,000." Clay offered.

"Who put up the investment for this mill, Clay?"

"Tom Landry and his quarterback, Danny White. They used to come down here to hunt big horn sheep."

We were standing on a hand hewn timber platform, looking down into a narrow crevasse that appeared to have been dug out with picks, shovels, chisels and hammers. The old jeep, with a broken spring was spouting steam and seemed to be catching its breath after bouncing up a steep and extremely rough narrow trail. The jeep driver was preparing the carbide lamps so the geologist could go down the incline and get his fresh samples.

Dick asked, "Why use lamps, Clay? We brought a flash light."

"If there is poor air down there, the lamp won't burn so bright. When the flame starts to falter it is telling you there is not enough oxygen and you had better get back outside." They used to use a canary for this warning, but it would be difficult to get a canary and its cage in this narrow crack. The flame was a better indicator of bad air than a dead canary anyway.

Samples are taken on an ore strike from the mineralized zones in the surrounding rock. Wilbur's geologist was going down to get the samples. Chipping samples from the rock, he would note the location of each and seal them in individual ore sacks. As they finished tying a rope around his waist and began to lower him down the shaft, I wondered should he have a Sign? No, I decided, this is his job. It is what he is paid for!

When performing an economic feasibility study on a project, your base premise must be correct. In the case of ore samplings from a mine, you can not take someone else's word for anything. I remembered what my Uncle Steve suggested years ago, a mine was a hole in the ground owned by a liar trying to sell it to a sucker.

The next morning we said adios, shook hands with the folks, and climbed into the plane. When Harvey turned the key, the engine refused to turn. Clay and George took turns turning the prop by hand until the engine fired up. This was a difficult process, as the turbocharger usually required high RPM's to start the engine. Maybe a burro is more secure for these parts of the world. At least with a hand full of oats, he will move.

Back in Tucson, I was telling Mary Martin about the adventure that we had. How we had discovered Dick had not connected a wire to the alternator properly resulting in failure to charge the batteries. Clay having to hand prop the engine to get it to start did not interest her so much as the question she had. "Why do you suppose the sun was in a perfect position to reflect the windshield of that airplane in the old building?"

"Who knows," I answered. "Guess you have to be lucky."

Mary was one of the people you meet once in a lifetime who

becomes a forever friend and cohort for the good or for bad. She was born in Pasadena, California in an orange grove. Her dad told her she was plucked from an orange tree. She was a direct descendant of Jefferson Davis. Her family had been quite wealthy, but lost everything in the depression of 1929.

Mary told me, "Once when I was little I had lots of pretty dolls. I got a hacksaw and cut the top off their ceramic heads. When my mama confronted me about it, I told her I wanted to be a brain surgeon and was practicing medicine."

Her family had eventually been forced to move to Arizona where her father took up the art of gold prospecting. Most of the boys from her high school graduation class had been serving on the battleship, Arizona. None of them came home.

One of the mines her family was working was just south of Tucson. She told the story of how a group of investors came out to the mine one day in a big black car to buy an interest in the mine. They were all wearing short pants. Her dad got his 30-30 rifle and ordered them off the property. He thought they were pretty dumb to come out in the cactus and rattlesnakes, dressed that way. They discovered later that one of the guys wearing the shorts was Joe Bonnano, the western Mafia boss.

The first time I saw Mary she was under the hood of her car changing the points in the distributor. She had called and left me a message to come and see her. "I'm looking for a Mary Martin, does she live here?"

Mary slid down from the fender. "You must be Mac," she said wiping the grease from her hands. "Let's sit under the tree and have a beer. We have a mutual problem." Wondering what that could be, I sat down and took a cold bottle from her. With no preamble, she said, "About two weeks ago your son took the old pick-up and left for California."

"School isn't out yet. Why did he do that?"

Mary smiled and said, "He took my seventeen year old daughter with him. You and I are going to be grandparents."

"Do you have any tequila?" I asked. Give the kid a Sign, I thought.

I was at Wilbur's lab helping prepare the ore samples that we had taken from the mine in Mexico. Each bag contained about a kilo of the quartz material.

The standard preparation procedure is to first crush the rock then pass the resulting granules through a pulverizer. The finely ground rock is then split into four parts. One part goes to the lab for analysis, one is saved for a backup check, another is saved for the mine owner. The fourth one is kept on file for future reference if there is ever a question on the final results. After each sample is prepared, the crusher and pulverizer are thoroughly cleaned with air so the succeeding samples will not be contaminated.

I gave the prepared ore samples to Wilbur and asked about the geologist. "He is a well known exploration geologist," Wilbur said. "Based in Canada now, he is always on call for the big mining concerns. He has a PhD in structural geology and is well versed in his field."

"If these samples are any good, what chance do we have in finding an investor?" I asked.

"Right now there's not much. The economy in Mexico is shaky and most companies are waiting to see what the peso will be worth."

"I'll need some financing for the Sierra Pinta, too. Maybe there's a way to tie them together."

"Let's look at these results first, then we'll look into that possibility."

We had fifteen samples to run. To get a hard average on several samples, the lowest value is thrown out and also the highest. The remaining thirteen gave an average of 15 grams or about one half ounce of gold in one ton of rock. Both the atomic absorption and the fire assaying procedures verified these results.

"Well," Wilbur said, "I guess you didn't waste the trip. This looks pretty good compared to most of the mines around here."

About this time two well-dressed businessmen were having a meeting. Harold, a short heavy set fellow, with baby faced pink cheeks and horn-rimmed glasses was discussing with his business advisor, Alphonse, a problem they had with a gold mine they had recently purchased.

"Our contract stipulates that we must start a mining program within 90 days," Harold remarked. "I saw in the Wall Street Journal this morning that the government has made the move to change the law on ownership of gold. It looks like we're cleared to go ahead with our plans."

"That's a big help. The next thing to do is to locate a mining engineer. I have a friend in Arizona who might know of someone," Alphonse said.

Since 1935 the US government had bought all the gold produced in the States. It was against the law for a citizen to own the precious metal, gold. However, anyone else in the world could buy and sell it. In effect United States citizens were relegated to second class world citizens. This was the main reason I was trying to develop gold operations in Mexico. At this time, gold was selling in the world at about three times the US price. The metals, silver and gold, would now be a commodity and traded in the markets like wheat, corn, pork bellies or coffee. In 1968 our government withdrew the gold guarantee on the US currency.

Mary and I had just taken groceries and supplies to the workers at the Sierra Pinta and were driving from Rocky Point back to Tucson. "How are you going to keep the mine going? Robert isn't going to help?" Mary asked.

"I don't know. I've exhausted the money I had to get this far. What do you suggest?"

"Maybe you should get a job," she replied. "One that pays, I mean..."

Women are like this. They never seem to understand that if you really concentrate on developing a project, there is no time left for a job. "Maybe you're right. Richard, has a project he's working on in Newport Beach, California. I should go see him."

I had not seen Richard for some time. We had first met in the University at Santa Barbara. He had a passion for sports cars, and we had attended a few rallies together. Later his experience with racing cars had resulted in his serving as editor for magazines such as Road and Track, and Car and Driver. The last time we had spent time together was when we had sailed his schooner to Ensenada during one of the celebrated yacht races. We won the race that day over Buddy Ebsen and his catamaran because we had broken a rule and used the engine instead of the wind. The last I had heard from Richard, he was helping to develop an external combustion engine in competition with Norman Lear and his steam car.

At the moment, we were sitting in the Rusty Pelican saloon drinking salty dogs. At this time he had just opened an office in Newport Beach, California, operating a company called Universal Silver Exchange. "Richard, what is this silver deal you're working on? You're supposed to be an engineer."

"The industrial demand for silver is rising, Mac. Just the requirements of the film business alone shows there will be production shortages. We think the price will start rising soon. I have a partner with an electrolytic refinery and he is making kilo bars of 99.0 % pure silver. I'm setting up now to market them to the public."

"Where do you get the silver to refine?" I asked.

"From used film and other industrial scrap."

"Who are you selling it to?"

"The public, the little guy. We think it is a better hedge for inflation. The average person can't afford the required 1000 ounce contract on the commodities market. I think he will buy the kilo bar. It's only 31 ounces."

"How much is a bar today?"

"One hundred four dollars and seventy five cents each!"

"How are you financing the operation?" I asked.

"I sold the boat for twenty thousand dollars and set up the office. Some of the money I'll use for advertising, the rest we have already used to buy the scrap silver."

"Richard, you know I've been working on the mine in Mexico. There is more silver than gold in the ore. I wonder if it would help you if I supplied the silver to you?"

"That's a good idea. We might have a supply problem by just using scrap."

"Who else is selling silver around here?"

"I've talked to one other guy in Beverly Hills. He's buying bars on the commodities market and reselling them to the public. Doing well, but doesn't want to deal in refined silver."

"Who are they?"

"U.S. Silver. They have a big market."

"Sounds interesting," I commented. "Let's have another salty dog."

Up in Alaska, a crew was working on an old gold dredge. It was an ancient bucket-line dredge moored in the Solomon River. These boats were low-draft barges, some large ones were three to four stories high. A chain arrangement, on a boom that lowered to the depth of the riverbed, had large buckets attached. The chain and buckets run around a large cogwheel, like a sprocket on a bicycle. The buckets pick up the rock from the bottom of the river or lake, and hopefully free gold and gold ore.

As the buckets pass over the top of the chain sprocket, they dump the material into bins, or hoppers. These hoppers distribute the incoming material to several riffle boxes. The riffles catch the heaviest material and the rest goes back into the river. This heavy material then flows through concentrating machinery that will vary in type depending on the material dredged. The concentrators separate the gold nuggets, if any, from the black sand. The black sand in the area around Nome, carries varying amounts of platinum and palladium, as well as tin.

The crew had just shut down the day's operation and Bob Slagle, the foreman, was helping with the cleaning of the riffle box. He had scooped out a bucket of the black sand concentrate and set it up on the railing of the boat. There was gold shining through the ebony crystals of sand. In his excitement at seeing all that gold, he let the bucket slip over the side. He grabbed the bail, but it was too much weight. It pulled him over the side. Bob said later, "There I was upside down in the river holding on to a bucket with about twenty thousand dollars in it and I didn't want to let go." Clambering back on board, he said to everyone laughing at him, "I think we've hit a good ore zone. I'm going to Los Angeles and find enough money to finish this project."

The opportunity here in Newport Beach seemed like a good way to create some capital to keep the mine working in Mexico. I stuck around a while with Richard learning about the refined silver business. He had made a cassette tape about the 'coming monetary crisis' and why it made sense to buy silver as an inflation hedge. It was narrated by Robert Preston, the author. We sent these tapes to every address on Lido Isle in Newport Beach. We did not get one response. No one believed there would be inflation. The working class was more worried about the growing gasoline shortage in the US.

Richard said, "We must be doing something wrong. Maybe we should try television." Obviously nobody was interested in owning silver, but we had decided to make him or her interested.

We both got a Sign! Like Fearless Fosdick we were going to save the world from the poisonous beans, even if it killed us.

Alphonse was on the phone with Wilbur, "Hi, Will. We have a good set up over here and need to hire a mining engineer. Could you recommend anyone?"

"Yeah, I know a guy. His name is Mac. I will get him to call you if I can find out where he is."

Harold said, "Glad to meet you, Mac. Wilbur Wright says you might be the man we are looking for." We were in the boardroom of nice looking offices on Rodeo Drive in Beverly Hills. It was a Saturday morning and no one else was around. The Brown Derby was only a few doors down the street. Harold and another man who he had introduced as his brother, Danny, were wearing short sleeved sport shirts. Their appearance was unfamiliar to me. I was used to working with guys in levis, red suspenders, and hard hats. They both looked awfully pale, and Harold appeared to be overweight for a fellow that looked to be in his late twenties.

"I've known Wilbur for four or five years," I said. "His company has been doing my assay work for a while. He told me you have a job opening. What does it entail?"

"We have a contract with an option to purchase a gold mine near Lake Tahoe and want to extend the tunnel to get to the main vein." Harold offered.

"Do you have any report or geological surveys on the place?"

"Yes, we have drilling reports and assay results as well as maps of the existing workings."

"What's your proposal as far as I am concerned?" I asked.

"If you can do the job," Danny said, "We'll give you a contract to oversee the project."

"Fill us in a little, Mac, what is your experience?" Harold asked.

"For the last eight years I've worked on several projects in the southwest and Mexico. I have experience in all phases of prospecting and exploration, development of milling procedures and underground mining. I've worked with McCullough helping to develop California City and their project at Lake Havasu, where we were doing deep water well exploration, using core drills. We analyzed the mineral in the rock at the same time. I spent several months with Heinrich's Exploration. We were all over the southwest doing induced polarization surveys for the big mining companies. We spent some time near Pecos, Texas, searching for underground caverns. Pecos is near where Judge Roy Bean held his courthouse in the saloon."

Have you fellows had any mining experience," I asked.

Harold chuckled and said, "No, we're just bankers." I felt at the time they were just two young guys trying to get into the mining industry. The pale look they had still puzzled me. I guessed they did not go out in the sun much. "Maybe we can use you, Mac. Let us think about this for a few days. Give me a call on Tuesday."

"Richard," I said, "This job offer from Harold and his brother could be what we need to keep going." We were back at the Salty Dog. The sloops and small schooners were slurping at their moorings and the halyards were drumming softly against the furled sails.

"Could be. You need money to send to Mexico and I don't have enough sales yet to really get started."

"You know, after all the years I've spent out in the desert and on top of mountains, this looks good to me. Beverly Hills has electric lights, running water," I remarked, taking a long drink of tequila.

Later that night I called Mary in Arizona. "Hi, sweetheart. May have found that job that you hinted at."

During the long years in the mining camps, Mary's mother had studied astrology. She was also a Rosicrucian. Mary had learned from her mother and was proficient at reading the stars. She answered, "It may sound good, but don't take it."

Something was square and something else was retrograde, but I thought, who cares. "That's too bad," I replied, "but I really need to get some money coming in."

"Don't take the job, Mac."

As I hung up the phone I sighed, "What a weird girl."

On Tuesday I drove up to the Beverly Hills Hotel to meet with Harold. Harold and another man were sitting at a table in the bar having a drink. Harold waved me over and said, "Mac, I want to introduce you to Mr. Morales. Alphonse, this is the mining engineer I told you about." Alphonse was Spanish, but he acted more like a prim Englishman.

"Glad to know you, Mac. Wilbur speaks highly of you." His handshake was limp and cold. I noticed his fingernails were newly done and highly polished. Never see that on a miner's hand. "I'm having Chivas Regal. Do you want a drink?" asked Alphonse.

"Chivas is fine, rocks, little water." I noticed Harold did not drink. Neither of them smoked.

I took out my pipe and tobacco. While filling my pipe, I observed both of them were well dressed but Morales' attire was impeccable. His hair was perfectly done and his movements were smooth and strong. Actually, he was too perfect. He seemed to be in total control of the discussions. I learned later, he had been an Olympics athlete and had competed in the pentathlon. He drove a hand made Jenson car that had never had anything more than a Kleenex in the ashtray. Harold said to Alphonse, "I have things organized and ready. How are you getting along with the contracts on the new business?"

"I just finished them this morning," Alphonse answered. "The down payments have been made. We're ready to go."

Harold said "Great! I need some action. We'll put Mac to work right away."

"Hold on a minute, guys. What are you offering, and what am I supposed to be working at?"

Alphonse answered, "We want you to come on board and oversee the operation of our new mine."

"That sounds interesting, what does it pay?" I could not tell them how tired I was of living alone and would take anything that would bring a change in my life.

"We'll give you an apartment here, get some good looking transportation and pay you one thousand dollars a week." Harold drawled. "How does that sound?"

After living several years away from modern civilization, with gila monsters, rattlesnakes and cactus, usually no electricity or running water, the picture they were offering looked pretty good to me. " I'll take it," I said, not remembering anything about the squares and retrogrades Mary had warned me about.

Whenever major policy shifts are made in government regulations, there are always 'loopholes in the law'. They are not really loopholes. They are opportunities for anyone who is paying attention to the changes. These opportunities are sometimes short-lived. At this time the government had made a shift in the way currency was handled. By taking precious metal backing away from the dollar and making it legal for the US citizen to own his own gold, loopholes and opportunities were found in abundance.

Loopholes are not necessarily right or wrong, but lie somewhere in between. This 'in between' is a gray area that is essentially the law of the prudent man. Many businesses operate, walking the thin edge of ethical and moral obligation. One can find this 'in between area' as large as a government or as small as the little gray haired schoolteacher reporting her income tax.

I asked Alphonse about their program. He answered, "Harold here has a new business start-up that is nation wide. His company will be selling gold to the public. You know about the law being changed on private ownership, don't you?" I nodded in response.

"In January this law comes into effect," Harold offered.

"What do you project for the volume of sales?" I asked.

"We know we can market a lot. We've already sold eight hundred ounces," Alphonse answered.

"So you bought the mine to supply this gold to the customer. How much gold is calculated to be in this mine?"

"We have the reports for you. Take them and read them. Part of your job is to tell us how much gold we can expect and how soon we can get it," said Harold.

This seemed weird. They have already sold their gold, but they do not seem to know how much they have to sell. I learned later that Harold had started selling before they had made a deal to buy the mine.

"What's the name of the mine?"

"The Gold Lake," said Harold, "near Tahoe."

"Tahoe is too far to drive from here. You should have an airplane."

"We're talking about that," said Alphonse.

"I know where we can find one," I returned. "The guy is having trouble making his payments."

"That's interesting. Maybe we should have a look at it." Harold and Alphonse turned away and began talking to each other in low voices. I looked around the room and wondered how many millions of dollars in business had been discussed here. The Beverly Hills Hotel was not the place where travelers just happen to book a room. It was not a cheap place. The vines and hanging plants were very relaxing. I wondered if Howard Hughes had ever sat in this chair.

Alphonse broke into my ruminating. "Mac, we are having a meeting in Palm Springs next weekend. Can you get the plane over there on Saturday? I'd like to take a look at it."

"I'll give them a call. This could be good timing," I answered.

Alphonse opened his briefcase. "Here are the reports on Gold Lake. Go over them this weekend and come to the office on Rodeo Monday afternoon."

"Okay," I answered, "see you then."

The reports on the mine were sketchy, but the drilling reports were legitimate. The assays were certified and showed high values where they had intersected the vein and the geologist was well known.

I was with Richard back in Newport Beach. "This mine looks like it has possibilities, Richard."

"Where is it, exactly, Mac?"

"Northwest of Tahoe, near Donner Pass."

It was late in the fall. This area of the Sierra Nevada had quite a history of heavy weather in the wintertime. "I better get up there before it snows," I said. "What are you working on?"

"I'm designing a chess set. We have a request from a savings and loan company in Salt Lake City to supply them with silver and gold chessmen," he said.

"Damn, that sounds expensive."

"Their idea is to sell each piece separately on a monthly payment plan, like the Franklin Mint does," Richard explained.

"What's the deal?"

"They want me to supply a chess set and they will sell to their customers, at the same time buying our silver and gold. They will give me a percentage on the mark up."

At the prevailing market price of metal, the chess set would have a value of about five thousand dollars. But since, the pieces were to be pure metal and required a very nice playing board, the total set had to be sold for twenty thousand dollars. The cost of refining and casting was very expensive.

"The Mormons understand the value of precious metals," I said. "But how can you get that much silver and gold to make the pieces?"

"That is a problem I have to do some more thinking about." he answered

On Monday afternoon, I arrived in Beverly Hills about two o'clock. The last time I had been here, I had not noticed the bronze plaque on the wall next to the door that announced U.S. Silver Corporation. Huh, that is interesting, I thought. Wonder if Richard knows about theses guys.

There were three guys walking out. They were dressed in some fancy clothes. They nodded as they passed me. They looked like used car sales men. On some level something more sinister registered in my brain, but I chose to ignore it.

Again, I found Harold in the boardroom. He was on the phone. "Sit down Mac. Be with you in a minute." I paid more attention to the room this time. The walls were all paneled in beautiful dark oak. The ceiling had large exposed beams and a dark blue Persian rug covered the pegged hardwood floor. The heavy drapes helped deaden the outdoor sounds.

"Sorry about that," he said as he hung up the phone. "What do you think about the mine?"

"It's interesting, but I'll have to go take a look."

"How much do you think it will cost to open up the old workings?" Harold wanted to know.

"I can only guess at this point. Probably fifty thousand, depending on machinery availability and the weather."

"That's not bad," he said. "That was Chicago on the phone. They just closed a large sale."

"What is your marketing plan, Harold?"

He leaned back in the chair and putting his hands behind his head, "The Company we have purchased, U.S. Silver Corporation, has sold silver bars to over three thousand people. Our sales force uses this mailing list for its contacts. These people have already bought silver and are a prime target for the gold sales."

They did not have any gold yet, but were taking in payments promising to deliver later. Essentially they were selling futures on a commodity. But this was one of those 'gray areas' that the present laws did not specifically say could not be done.

Changing the subject, Harold said, "We've cleaned an office for you. Come upstairs and take a look." The upstairs office had windows overlooking an alley. I could see one of the doors across the alleyway had 'Gucci' stenciled on it. "You can park here behind the building." Harold pointed out. "We'll have a car for you by next week."

So, I am in business I thought.

Back at the Rusty Pelican, Richard and I were comparing developments. I had upgraded to Chivas and rocks. Tequila and beer were the poor miner's drinks. "I just completed a sale for twenty thousand dollars in silver ingots," Richard said.

"Hey, that's great! Who bought them?"

"A little old lady from Pasadena," he said. "Mrs. Brown, a widow, she wants me to hold it in trust for her and resell it when the price goes up."

"So now you are a silver trader," I laughed.

"Before I forget, there was an inquiry today from a guy who says he has a mine in Alaska. Wants to find an investor to set up a process and refining facility."

"What kind of mine? Did he mention?"

"Yes, he has an old dredge to put back to work. It's up near Nome."

"Interesting. Better call him," I said. These dredges had been used since the California gold rush days. They were very efficient in handling large volumes of material that were under water. There had been dredges operating around Sacramento and in Colorado. When the gold had run out, some of these boats had been taken apart and moved to other areas. Some had gone to South America and this one to Alaska.

"From what I can determine in studying the Gold Lake project, it will be a long shot if there is enough metal to supply Harold."

"Maybe we can interest them in putting some money into the Nome project." Richard suggested.

"That is a possibility. What's his name?" I asked.

"Bob Slagle," Richard said. "Lives in Glendale."

Bob Slagle was an average fellow, short and heavy set. He told me he worked for the Jet Propulsion Lab in Pasadena as a maintenance mechanic. I learned later he worked as a janitor at JPL. He lived in a mobile home with a wife, but I never met her. Bob said he had several years experience working in Alaska. I saw no reason not to believe him.

"Mac," Slagle said, "the operating season is only three months in that area. I want to make as much concentrate as possible and ship it down here for processing. We need enough money to do repairs on the boat and operate for the season."

"I may have a way to do this. Give me a couple of weeks to work on it."

"Let me know as soon as you can, Mac. I want to quit my job at JPL.

The meeting at Palm Springs took place as planned. The pilot, Harvey, had flown Dick's Cessna 206, arriving about eleven. Mary Martin, Clay, and Dick, the owner of the plane, were passengers. Mary and I were sitting by the pool at the Tropicana Hotel. Dick was inside talking with Harold and Alphonse about the sale of his airplane.

"I sure do like the feeling of Mortimer," Mary said.

"So you are calling the plane Mortimer, too. How long did the trip take?"

"About three hours. We had a little headwind."

"How does Dick feel about selling the plane?"

"He's pissed, but doesn't have much choice. The bank is trying to take it back."

Harvey came over and sat down. He had a beer. "What have you been up to, Harvey?"

"Have been doing some instruction and had a charter last week, but it's really slow."

"If these guys get together with a deal on the plane, I will be able to use you as our pilot."

"There could be a problem, Mac. Dick says he won't give it to them if he can't go on all the flights."

"That's ridiculous. How could that work?" I said.

"It wouldn't, but that's what he wants. He's in love with that airplane."

Harold's brother, Danny, came out of the hotel and approached our table. "Alphonse wants to see you. He's in room 666."

"Thanks, Danny. Excuse me a minute guys. I'll go see what has transpired.

Alphonse's room was a suite. He and Harold were relaxing in the overstuffed chairs.

"Come in, Mac," Alphonse invited. "I bought the plane from Dick. He'll get all the papers to me by next week."

"Okay, now what's next?"

"We just closed a finance deal here at the hotel. We'll cut you some operating money from that on Monday," Harold said.

"Get ready to go to Tahoe. We're ready to start operations." Alphonse said.

"Sounds good to me," I answered. "See you next week."

The apartment in Beverly Hills had been part of the deal when Harold purchased U.S. Silver. It was located in a full security building. The address was Number One, Doheney Drive. I had been given a key to the parking area and been introduced to the armed guard at the door. Mary had taken a few days off her job and was resting on the couch.

"Well, Mac, you have come up in the world. But are you sure about these people? I don't like any of them. They appear to be pretty slick to me."

"Yeah, I know, but they seem to have money."

"Money isn't always what it appears," Mary said.

"Well, we'll see. Let's have a drink and celebrate. This is going to be a big deal!"

"Sure it is," she said. "I'll have a straight shot and a beer."

Back in Palm Springs, Harold, Alphonse and another man were finishing up their business. "I think we can use that Mac. He'll fit into the scheme of things very well," the man said. Earlier this guy had been out of sight in the other room when I had discussed with Alphonse and Harold about going to work for them. I did not know he was there.

"I think he is perfect for the business," Alphonse said.

On Monday morning I was all charged with excitement. Dressed in a suit and tie, carrying my worn out briefcase, I slipped up the stairs to the new office. There was a tall, well-built guy there. The style of his clothes and the Italian shoes reminded me of the movies about Al Capone. He was about 6'4" and moved like a cat. "Well, good morning," I said in surprise. "Who are you?"

"I'm Roger," he answered holding out a big paw. "Harold asked me to show you around some."

"Gee, that's nice. Where are we going?"

"We have to pick up your car this morning. You can ride with me and bring it back." Roger had a big tan Mercedes sedan. We turned off Rodeo Drive onto Wilshire Boulevard, he said, "Guess we're going to be working together."

"Oh?" I replied.

"Yeah, Harold said he didn't have a lot for me to do right now. Said I'm supposed to learn something from you about the mine and help you all I can."

"Do you know anything about mining?" I asked.

"Nope never saw a mine."

"Where did you go to school, Roger?"

"Didn't get much school," he answered.

"Hmm, where are you from, around the LA area?"

"Croatia," he said. This will be interesting I thought. I wondered if this big teddy bear could spell 'mine'. "Al Morales has a leasing company. He's got two new Plymouth sedans for us. One for you and one for Harold," Roger offered.

"You didn't go see the Gold Lake mine with Harold, did you?" I asked.

"No, and Harold didn't see it either. He made the deal at the hotel in Lake Tahoe. Was a fun trip."

We picked up the car in Commerce City and I drove it back. Back at the office, Roger said, "Let's grab some lunch. There's a nice little bar down the street."

I came in the door of the apartment, after talking my way through the new security guard. Mary was reading. "How was your day at the office?" she asked.

"I never saw it. Met a guy that works for Harold, a Croatian. Reminds me of a big teddy bear."

"Was he carrying a gun?" she asked.

I had not thought of that and wondered why Mary would ask such a thing.

Stephan Riess was kind of a relative. His wife was my mother's first cousin. He was the force that had pushed me into the mining industry. Steve had come to the US from Germany after the first

World War. His father had been a general under the Kaiser. His major studies were in geochemistry. With that knowledge, he had been able to land a job with Herbert Hoover as a mining engineer. During World War II, he had worked for the government exploring for strategic minerals.

After I graduated from the university, I traveled with Steve to various mines and exploration projects around the Southwest. Steve's experience in Germany during the period of high inflation and getting through the economic crash of '29 in the US was real history in the flesh.

He told me that the final ratio of inflation in Germany, after the First World War, had been one trillion to one. He had known the Minister of Finance and one day asked him, "Minister, will we ever run out of money?"

"No, but we are going to run out of paper to print it," the Minister had answered.

Steve's philosophy was that without a base for every currency, the people in a country could never have real wealth. "Mac, I want you to remember this story. When I was born, the family put enough money in trust for my education. When I was twenty, I received a letter from the bank requesting that I come get my money. They didn't want the account any longer."

"That was a help. You had a lot of cash then."

"Oh, sure. I walked across the street and was able to buy a box of cigars with the whole lot. With gold and silver backing the currency this could not have happened."

I remembered a story my dad had told me several years before. A German who was working in his shop had told how he had been a carpenter in Dusseldorf. Every day when he left work he took a few nails home with him. When he had filled a keg with nails, he traded it in for a boat ticket to the United States. The iron nails had more value to them than paper money. These stories and experiences with Steve had convinced me that precious metals were something that I should learn about.

The program that Harold had developed had a major problem. They were working on selling bars of metal to the public. By buying a mine they were not mining bars. They were mining ore. This ore had to be processed and refined before you could pour an ingot. Harold and his group had not the foggiest notion of this part. It seemed to me that they did not care. Their job was selling and they were very good at that.

"Mary, we can put in the concentrating plant for the mine in Tahoe, but we have no way to do the final refining of the metals."

"Can't you use Richard's partners for that?" she asked.

"No, we can't. They are set up for electrolytic refining of silver only. They aren't processing concentrates from any ore. When you were working with your dad, you concentrated the heavy mineral particles and sent them to the smelter. The smelter melted everything up and poured an ingot. To refine the final metal that the smelter wants, they take the ingot into solution by electrolysis and then they extract only the specific metal ion from the solution. With the proper control, very high percentages of one metal can be achieved. This is called electro-winning. Silver is one of the easiest metals to refine in this way. Gold is treated the same, but usually the problems in extraction are much more complex."

"Okay. Why not call your Uncle Steve? Didn't you tell me he used to work in this field?"

"What a great idea. I will call tonight and see if he is around."

Uncle Steve was now working on developing primary water wells. Using the extensive knowledge he had in earth chemistry, he had worked out a method to locate drilling sites from the basic geology of the deep-seated rocks and the crystalline structure of the minerals.

"Steve, I have a project where we will be required to build a refinery. Can you help me on this?" I had called him at home. I was lucky as he had just returned from a project in Mexico.

"Jim Scott is your man. He's working with a chemist on platinum recovery. They are down near Saugas, California. He and Monty are developing a lab there," Steve said. "Here is the number." I wrote down the number to call Jim. Jim Scott had worked with Steve for twenty years as his field engineer. I had worked for him on the drilling program at California City.

"Jim's not drilling now?" I asked.

"No, we're between projects, so I sent him down there to learn about the separation of the platinum groups," Steve answered.

Back at the office, I was preparing to go to Tahoe. I saw Harold wandering around downstairs. "Hi, Harold, how's it going?"

"Very good! We're getting orders every day."

"I need the number of the guy at Gold Lake. I'm about ready to go up there."

"I just talked to him, Mac. He said they had a snowstorm last night. You'll have to wait a day or so until it melts."

"Okay. But we're going to run out of time."

"Wait a few days. They will call when the snow melts." Behind Harold there was a large walk-in vault. The door was open just wide enough that I could see a stack of large silver bars inside.

Mary had a job at a private school in Tucson. She was the maintenance person and the bus driver. After she had started work there, the administration had fired two other people. She was doing the work of three now, and of course, was not getting paid any extra.

"My holiday is over, Mac. I have to go back."

"I have a car now. Why don't you drive Merlin, the pickup?

"That's a good plan. When you return, you'll have him to travel back to the Sierra Pinta. I will leave early in the morning. You be careful with this group and don't take any more wooden nickels," she admonished.

"Roger, I'm driving out to Saugas to talk to a friend, Jim Scott. Want to come along? You can learn something about mining."

"Much better than setting around here," he said. "Let's go."

We took the ramp to the 405 Freeway North and headed north on I-5. "How long have you worked for Harold?"

"I just started this job. I worked for him about five years ago on another project," he said.

"What is Harold's last name? No one has ever mentioned it."

"Goldstein," he said.

"That's good. Goldstein, in the gold business."

We found Jim Scott at work at the lab. Jim was a tall raw-boned man who had spent his life in the outdoors in the oil and mining business. He used to say his main claim to fame was that he was on the only plane to ever crash on landing in San Francisco Bay. It was a military transport bringing soldiers back from the Philippines during World War II. Uncle Steve always said, "Jim is so conscientious about his work that he would polish the rail on a sinking ship."

"Hi, Jim, long time no see," I greeted, shaking hands. "This little guy with me is Roger."

"Glad to meet you, Roger. Come on in. Have a stool."

The air inside was heavy with chlorine and nitric smells. A funny looking little round man was stirring a pot of acid with a wooden spoon. Now here is the original Merlin, sans beard, I thought. He really did look like an elf.

Jim said, "This is Monty Lynch."

"Hi there, Monty. Glad to meet you." He was a fussy little guy and talked with his hands a lot. He waved them around proclaiming he had to get back to his pots.

Roger complained about the bad atmosphere and said he would wait outside. "It's no worse than the LA smog," I said. "Go ahead. I'll be out in a while. What are you working on, Jim?"

"Monty has developed a chemical process to separate the platinum group metals from silver and gold," he said.

"Wow, does it work on black sand concentrate?"

"Yes, it does. That's exactly what the process was developed for."

"How does it work?"

"Basically, it's pretty simple. We use hot concentrated hydrochloric acid to oxidize the ferric iron. The precious metal ions are released and go into suspension in the liquid. Monty is adding pure aluminum beads to the filtered acid. The aluminum acts as a catalyst and the platinum metals and gold collect on the aluminum. From there we go to the hydrogen smelting process to pour an ingot," Jim explained.

"Then you can use electrolytic plating to separate the gold and platinum. Is that right?" I asked.

"Exactly. The plating process is well known. The trick is to get the precious metals ions to release from the iron," Jim said.

"Man, that sounds interesting."

"Want a Pepsi?" Jim asked. Jim did not drink alcohol. "Helps clean the acid from your sinuses."

The building complex here was originally a chemical handling facility. There were several large metal buildings. Each had high voltage power with large transformers. This was a perfect location for a complete concentrate processing plant.

Outside in the fresh air I asked Jim, "What kind of a deal do you have here? Who owns this property?"

"A Mormon Bishop owns it. Monty has an agreement with him to purchase everything here. The owner's name is Jim Stokes."

"How are you going to finance this project?"

"We're looking for someone now. Got any ideas?

"Maybe I do. Where are you planning to get the ore to process?"

"I've been working on a mining property near Susanville that looks very good," Jim answered. "It's an old workings, just north of Reno in the Sierra Nevada range near the Lassen National Forest."

This bit of news was interesting. Susanville had a history of some large high-grade gold and silver mines. The famous Comstock silver mines were not too far south and on a direct line it was quite close to the Gold Lake operation we were supposed to work on.

The first settler of the area was Isaac Roop, who established the Republic of Nataqua. This area joined the Nevada Territory. Mr. Roop was the first Provisional Governor of Nevada and later a US Senator. In the 1860's, California surveyed the area and discovered it to be part of California. The local citizens did not want to accept this change and fought back. After a day long gunfight, they gave up and the town became part of California. It was named Susanville after Roop's only daughter, Susan. The town was incorporated in 1900. Today there are about thirty four thousand people living in the city.

"And you need financing for that too, I imagine," I remarked.

Jim chuckled, "You know enough about mining by now to

know you always need money."

"I may have a way to do this."

"We're open for suggestions. Who's Roger?"

"Roger is my bodyguard," I answered jokingly, thinking of Mary's question about the gun.

In the car on the way back I said, "That was worth the trip, Roger. Jim and Monty have a refining set-up started that we can use."

"I didn't like all of those red fumes boiling out of the glass pots. My throat still hurts."

"When you are in the lab, you have to learn not to breathe."

"I didn't know it took so much to make gold."

"That's only a small part. Jim told me one thing they need for the process is a method to grind the concentrate into a fine powder before they add it to the acid."

"Do you know a way to do that?

"In Tucson I saw a machine that was built just for that. An old man by the name of Roy Burkett invented it. I don't know him, have only talked to him on the phone, but I'll check with him as soon as I can. I think he has a unit he's not using."

A news bulletin broke into the car radio station. The announcer said, "There are heavy snows in the Sierras. Travelers are advised that Donner Pass will be closed until further notice…"

We fought our way through the traffic on the 405 and the Ventura Freeway interchange. We arrived at the office late in the afternoon. No one was there except Alphonse Morales. He was coming out of the vault. When he saw us he quickly closed the door.

"Hi, guys. Where have you been?" Alphonse asked.

"We went to see a friend of mine about a possible refining facility," I answered.

"That's good news. I'm on the way down to the Playboy Club on Wilshire Boulevard. Follow me, we'll have a drink and talk about it."

"Do you have a key?" Roger asked.

"Everyone in Beverly Hills has a key."

I had been in a Playboy Club in London a couple of years

before. This one was different. There was no gambling visible. The bunnies were the same though. Too bad the bunnies at the mine in Mexico do not look like this. Our kind of bunny was our cook with the gold tooth. "Al, we heard on the radio that there is a heavy snow storm in the mountains near the Tahoe mine," I remarked.

"No kidding? That could be bad for us."

"Yes, considering you are promising delivery of gold to your customers by spring."

The storm had laid down sixteen feet of snow over a four-day period. There had not been a storm on record as bad since the Donner Party got trapped the winter of 1846-47.

Monday morning we were back in the office. "This early snow storm is presenting us with a problem. They called from Tahoe and said not even the snow cats can get through," Harold said.

"And traffic both ways over Donner Pass is stopped," added Roger.

"Mac, you know where some other mines are, don't you?"

"We have the two properties in Mexico, the Sierra Pinta and Clay's mine can be put on line. Also, I have a lead on getting Black Sands from Alaska. I can call the guy from Alaska if he's still in town."

"Why don't you get Clay to come over and we can structure a deal."

"I'll try to find him. We'll need more money to start these projects, Harold."

"We'll set up some accounts for this. Don't worry about that. Just get to work."

"Yes sir," I replied.

Bob Slagle answered the phone. "Say, Bob, this is Mac, we have a deal cooking here. I think we can work with you on the dredge up in Nome. Can you get the owner to come down here? Harold will draw up a contract and start organizing how to get the material to us by spring."

"Sounds good, Mac. I think he'll come down. Maybe I can quit my job at JPL, huh?"

"Looks like it. I'm going to need a lot of help. I have to get to Mexico. See if you can set this up for about two weeks from now."

I got Jim on the phone, "Jim, how soon could you get the mine in Susanville into operation?"

"If we get the funds right away, we could get some concentrate delivered in about three months, I'll try to get a message to Charley at the mine."

"Looks like we are in business, Jim, hope you can alert him soon. I'm on the way to Mexico. See you in a couple of weeks."

We were in a small hotel in Chihuahua. Clay had made an agreement with Harold. Harold would supply the cash to buy machines and get the Sierra Madre Mine, at George's ranch, into production.

In the mountains, near the small village of Pilar, there had been a gold operation run by a company called Earth Resources. All of the processing machinery that we would need to get started was there, according to Clay. Clay made arrangements to look at it. I had taken the advice of George about not flying into an area you did not know. The day before I had flown down with an air taxi to watch and learn how to approach this landing strip. Everyone, who knew it, said it was tricky.

"Okay crew, up and at 'em. Time to get out of here. It will be daylight soon." Clay loved to be the straw boss. They had done the preflight checks and filled the fuel tanks on Mortimer. I rode right seat to point out the problems to Harvey who was again our pilot.

"This is an interesting place, guys," I said. "But try not to pee in your pants when we go in there."

Flying early in the morning is generally smooth. This part of Mexico is all rock, not many trees. The many colors caused by oxidization of the minerals were beautiful. The varying reds from the iron and yellow oxide created by the sulfide breakdown could never be matched by the artist's brush.

We were approaching the higher mountains. "I think that's the canyon we want, Harvey," I said pointing off to the southwest. We banked slightly and approached a craggy steep sided opening in the Sierra Madre mountain range.

"How far is it?" asked Harvey.

"About twenty five minutes. We will see a column of rock," I explained. "Looks like a big chimney."

The canyon walls were moving closer to the wing tips. We were following a river and had to find another smaller canyon coming from the west. "There's our column. Harvey, go round the left side. If I'm right, there will be a break in the wall to the west."

If we missed the entry we would be in trouble because further in, the canyon narrowed as the mountains grew bigger. There would be no room to turn around. We turned slightly to the east and passed near the chimney.

"Okay, we're on track. Turn west now," I instructed. A sharp bank to the right put us in line with the side canyon. "See the little lake up there?"

"I see it," Harvey answered. He throttled back a little and I lowered the flaps about 10 degrees. This was tricky. The landing strip, after the first 300 feet, went up the side of the mountain at about a forty-five degree angle.

"We have to get up that incline, Harvey. As soon as we're down, I'll dump the flaps. Keep the ground speed about ninety. We need enough power to get to the top."

Harvey increased the pitch of the prop. There was a small flat spot at the top of the hill. Actually, it was just enough room to turn Mortimer around on. A larger plane would not fit.

The interesting part of this landing procedure after touching the ground, you had to fly up the side of the mountain. Mortimer put his gear on the ground just past the lake at 110 mph, slowed to the prescribed ninety to begin his climb. If the pilot miscalculates here, two things can happen. One, with too much power you hit solid rock at the top. Or if not enough power, the aircraft stops on the slope and slides backward down the hill. Either way the plane does not ever leave again and chances are the passengers do not either.

This approach had all of Harvey's attention as well as everyone else riding this roller coaster. Harvey managed his job perfectly. We popped over the edge of the flat spot and, through the spinning prop, saw twenty Indians squatting next to the mountain. I was not sure who had the greatest thrill, the injuns or us. Each of

Mortimer's six seats had pucker marks.

Who got the Sign? Everyone, including the Indians.

There was a natural cave in the limestone. The Earth Resource engineer had enlarged it some and placed all their processing equipment underground. Everything, including a large caterpillar generator set that was in good condition. With a little TLC, it would be ready to operate.

"This looks good to me. We can move all of it over to our place. Shouldn't take more than three months," Clay offered.

"What a good break for us, Clay. We can save a lot of time and money by getting this equipment. Won't have to go through the red tape of shipping across the border," I said.

"Let's go see Pedro," Clay answered.

Pedro Gonzales was the Mayor of Pilar. He owned a small general store. The campesinos traded him gold and silver for flour, coffee, shoes or whatever Pedro could pack in. He also had some small mining operations working.

"Clay tells me you have a mining business in California, Mac," Pedro said, as he handed me a can of cerveza Tecate.

"Yes, we do, Pedro. We'll have a refinery working soon."

"I have a bar of gold that must be refined. Would you be able to buy it?"

"What price?" I asked.

"Fifty percent of the gold value. You can have the silver and copper," he said.

"We couldn't buy it before we assay for the gold value," I explained. "What does the bar weigh?"

"About seven kilos, fourteen to fifteen pounds," he said.

I quickly calculated that if the bar was in fact fifty percent gold, it would be worth fifteen to sixteen thousand dollars. "We can sell the gold, Pedro, but we can't buy your bar until we assay it."

"You take the bar with you. Pay me when you find the value."

"Pedro, why would you trust us with your gold?"

"I don't, Senor. This is my daughter, Elodia. She will go with you and watch the oro."

Elodia was a striking girl. She looked at me with dark eyes full of mischief and smiled. I could only take a sip of the Tecate and blink.

The next thrill was slipping over the edge of the landing pad where we had parked Mortimer. The incline was steep enough that we were airborne within 100 feet. An easy bank to the north sent us down the canyon and on the way back to Nogales, Arizona.

Flying back into the US was much simpler than flying out. There were no laws that said we could not go directly to an American Port of Entry. The procedure to cross the border was fairly easy. The regulations said to call the customs and immigration by radio at least one-half hour before arrival. We were getting in after dark and this station did not maintain personnel 24 hours a day. But they would come out to the field and check us in by request. We knew most of the guys who worked on this station and as a rule they would just do a cursory search of our gear and wish us a safe journey.

This time, however, was a little different. We were bringing into the country a beautiful young Indian girl and a bar of gold bullion. While it wasn't illegal to bring in gold we were supposed to declare it. This declaration was for the Internal Revenue Service. Obviously this was a lot of red tape and paperwork that we didn't want to do and neither did the customs officials. The gold bar was in the glove box in the instrument panel in plain sight if they opened the box.

A black haired Indian girl wearing buckskin was more difficult to overlook than a bar of gold. It took more time than usual to clear the checkpoint. Elodia had proper identification and was able to acquire a twenty-one day entry visa. I had formulated a story about why we were in possession of an ingot and what we intended to do with it. The official never spotted it, so we kept quiet.

Later we landed at Tucson International and parked the plane where we had found it. Dick will never know he missed another great adventure.

The next day I was able to locate Roy Burkett in Tucson at the Gila County sewage plant. He was directing a welder on what needed to be added to his machine. This was my first face to face meeting with Roy. Roy had been a lobbyist for a large oil company in the past and somehow had acquired the patents on these machines he was working on here in the county yard. Later I was to learn he was not the inventor. He was really a politician and promoter.

"Mr. Burkett, I understand these machines will pulverize rock, is that correct?"

"My patent designs were originally based on the need of the mining industry for finely ground ores," Roy answered.

The unit the men were attending to was about ten feet high and looked to be about five to six feet in diameter. "What are you using this green machine for?" I asked.

"This machine is designed to shred household garbage. We are just about ready to test it for the county sanitation department."

Not too interested in hearing about garbage, I asked, " Where is the machine that you use for rock?"

"Right over there, by the methane plant. This is a mid size unit. I have been testing organic wastes with it. Does quite well blending corncobs, straw and manure. Makes a high class compost," Roy explained.

"We're putting together a processing plant in California and need a way to reduce rock particles to a very small size. Is this one for sale?"

"Yes, it is. If you have a project to put it to work soon, I'll make you a good deal. Tomorrow I am giving a demonstration to a group from Colorado. They want to see the unit pulverize their silica. You are welcome to watch if you like."

"You bet I would. We'll be here."

Elodia and I attended the demonstration. I got a business card from a man from Colorado who was there to observe the demonstration. He asked me to call him if I was ever near Denver. We then climbed in the truck and drove to Beverly Hills.

The lab in Saugas impressed Elodia, but not as much as the sight of Los Angeles. She had never been near a large US city, let alone anything like LA. As we entered the lab, we saw Monty stirring one of his pots. Across the room, Jim was gazing into a steromicroscope. We carried the ingot to the bench and dropped it with a loud thud. "Hey Jim, come look and see what I found," I called.

Jim peered over his glasses and said, "What? The girl or the bar of metal?"

"Both!" I grinned. "Elodia, this is Jim Scott. He is the one who will determine the value of the bar."

I was explaining the new gold find to Roger when Harold stuck his head in the door. "Hi Mac. Come with me, will you? I want to show you something."

We walked down the street, toward the Brown Derby. "We made a deal on this place. You and Roger will move your office down here."

"Why, Harold? Won't this place add more expenses?" Harold ignored that. "You have an appointment with the attorney this afternoon. We have set up a new company to handle the mining properties and you guys are the officers."

"What company?"

"It's called Gold Lake Mining Corporation."

That afternoon I signed the incorporation papers as President of Gold Lake and Roger as Vice President. "Let's get a drink and toast to this," Roger said as we walked out of the building.

"This development calls for two drinks. Let's go."

Every one in the circle working on this mining project with me was what one would call an 'old timer'. They had come up in an era where a certain code of behavior prevailed. The prudent man had a basic sense of what was right. He knew the consequences of doing wrong. If you stole a horse you were hung. If you jumped another's claim, you were shot. If you were caught in a storm, you could use a cabin and eat the food. You left a thanks note with money to pay for it. Newly chopped wood replaced what had been burned. Deals were made and consummated on handshakes. The mountain-man's word was his religion. What the old timer did not understand, or did not pay attention to, was that the new breed, the new generation, did not operate within this moral code.

Jim Scott's favorite saying when someone committed an indiscretion was, "They're young yet, they will learn." But in my way of thinking the cost to the innocent was very high. I remembered an adage: "Confucius: A gentleman takes as much trouble to discover what is right as the lesser man takes to discover what will pay."

The world economy was in an inflationary mode. The prices of both gold and silver were edging up. Daily, more and more articles were appearing in the papers and investment magazines relating to the rise in the price of precious metals. Historically as inflation rises the price of metals rise also. Richard had built an actuarial table based on this information. According to the available information, the price of silver was projected to rise by the early eighties, to three thousand dollars per ounce. For the first time in history the value of the British pound was allowed to float, mainly caused by the high rate of inflation in Europe.

"If silver is supposed to go up to three thousand an ounce and gold is supposed to be worth sixteen times the value of silver, then our gold mines are priceless," I said to Richard. We were caught-up in the excitement of the moment. Richard had bought another sailboat based on the presumption that we were all soon to be very wealthy.

The three of us were sitting on the fantail enjoying the sunshine. We had a weekend of relaxation ahead of us.

"You're right, Mac. When the mines start producing we'll be in a perfect position to take advantage of these conditions. What do you have in the tobacco can?" Richard asked.

Elodia had a one pound tobacco can tucked under her arm. "Jim Scott reduced the gold ingot to chips in order to assay it better. The can is full of gold!"

"Damn, that's heavy," he said as he took the can from Elodia.

I smiled and said, "There is about sixty ounces in there. Now we have to sell some of it so we can pay Elodia's father."

We walked over to the Salty Dog Saloon and just for the hell of it we took the gold can with us. The waitress said, "What's in the can, guys."

"The can is full of gold," we answered.

"Oh, sure," she said. "Do I look that stupid?" Once in a while, no security is the best security.

"What's next, Mac?" Richard asked.

"Slagle is bringing the guy from Alaska in next week. Harold wants to buy his dredge. Also we are getting a purchase contract ready to take over the Saugas property for the refinery. The Mormon Bishop said he would sell it to us. That is about all of the major stuff. What are you working on?"

"I'm having the chess pieces cast for the deal in Salt Lake City. They will be on display for sale next month," Richard said.

"Where did you get the metal to cast?"

"I worked out a way," he said and left it at that.

"Roger, it looks like we're getting things tied together," I said. Slagle and his man from Nome had just left after signing the agreements with Harold to purchase the gold dredge on the Solomon River. "I feel good about all of this, Mac. Putting the mines together with the refinery makes a nice package," Roger said.

"It does, Roger, and the supply of bullion from Elodia's father doesn't hurt a bit. We're in business. Looks like we've got it made. Let's go out and see the Bishop. He is interested in buying some of Elodia's gold."

"All right. We need to get away from the smog for awhile."

Jim Stokes, the Mormon Bishop, was a no-nonsense fellow. He had flown B24's and 17's in the North African campaign during World War II and was now a very successful businessman. "My wife and I have been saving our pennies," Jim said. "We believe in gold and have decided to buy what you have there."

The phrase saving our pennies, had been around for some time, but who actually saved pennies? Jim had! We weighed out eight thousand dollars in pennies. Who can imagine that many pennies in one pile?

"Let's go to the bank and trade these in," he chuckled. "I'll give you a certified check for the balance."

"Why do you have so many pennies, Jim?"

"For the copper value. The market price of copper will eventually make the cent more valuable as metal, than its value as a currency," he said.

Soon the government would add zinc to our penny and some day we may see the penny discontinued.

I took Elodia to the LAX airport and put her on a Mexicana flight to Hermosillo. "Tell your father to keep mining. We'll be down to buy some more. This is a good deal for us all." She smiled and disappeared into a Boeing 727.

I was in the office working out the logistics for our projects. I had sent Slagle to Arizona to bring back a truck and machinery we could use in Saugas. Jim had gone to Susanville to help organize the beginnings of production there, and Clay was off to Mexico with Mortimer helping George get their Sierra Madre mine at the ranch in working order.

Harold came in the door and said, "Mac, I need you to do something."

"Oh, what is it?"

"I want you to go to Mexico for me. I can't go."

"What's the deal, Harold?"

"We have a contact in Chiapas who has some gold bars to sell to us. I want you to go look at them," he said.

The state of Chiapas is located in southern Mexico on the border with Guatemala. I had been down there a few years back surveying a silver mine. "I know a little bit about Chiapas. What do you have in mind?"

"I've arranged for a DC3 to fly you down on Saturday morning. If you think the gold is okay, then bring it back with you."

"How much are we talking about?"

"There is reportedly a lot. We have been offered a good price. We can make a killing with it."

"How much are you paying?"

"A million dollars," he said. "All you have to do is check it out, pay them the million, load it on the plane and bring it back."

I had done some nutty things in the past, but nothing that approached a scheme such as this. I slowly picked up my pipe and started to fill it while deciding how to answer. Lighting the pipe I said, "When I was down there I met a girl whose grandfather was Yamamoto, the Japanese admiral of Pearl Harbor fame. I didn't see him, but he's there. Other people told me some of the old

Nazi hierarchy were living nearby. The insurgent guerrillas use the area for a safe haven each time they blow up something in Guatemala and the Mexican army chiefs seem to be all bad-news kind of guys."

"Yeah, but we have connections. The police will protect us," Harold offered.

"Let me understand this better, Harold. You want me to take a DC3 into the jungle, sample their gold, if there is any, pay a million dollars in cash and count it out on a log, then load up several bars of gold and quietly disappear into the great blue yonder. Is that correct?"

"That's right. We'll make a lot of money."

"The airplane and I will be protected by the Mexican police and their automatic weapons, right?"

"Oh, yes, we'll pay them for the protection."

I exhaled a large cloud of tobacco smoke and said, "Harold, I think you need a vacation. You take the plane on Saturday. I'm sure I can find something better to do."

Harold shrugged, put his hands in his pockets and, as if in a dream, walked out the door. He did not take the vacation.

I recognized the Sign that day. It was a big one and probably read R.I.P. with 'P' standing for pieces.

Roger and I were sitting in an office in Atlanta. Harold had summoned us to come down here to explain the background on the mining operation to his brother Danny. There was another fellow in the meeting who had been introduced to me as Bert. Roger seemed to know him from the past.

"So you have a supply line set up to help with our gold sales?" Danny asked.

"Yes, we do, all these mines have been tested and proven to meet the potential tonnage of ore required. We are starting the initial work on the higher assayed values of gold to catch up to the demand."

"What else do you need." Bert asked.

"We're going to need more cash. We are working on a crash program, to gear up production as quickly as possible."

"We can arrange that. You can have the money right away. Perhaps we can get you some by Thursday." Danny said.

"Come with me and meet some of the guys." Bert said. "They need a pep talk."

Danny, unknown to Roger and me, had taped the conversation of this meeting. We entered a large room where there were ten or twelve people working. There were a couple of women present also. Each person had a telephone. They all looked like used car salesmen.

Bert said, "Okay people, shut off the phones for a few minutes. This is Mac, our mining engineer from California. He'll give you an update on the progress of the operations out there."

Not really understanding the whole picture here, I blinked and asked, "What do you want to know?"

To a man, the basic question was, "Are there really mines? Do they actually have gold in them?"

"Of course there is gold. Why are you questioning that?" I asked haltingly.

"The last company we worked for did not have the product we were selling," a salesman said.

"That's right and we were afraid this was happening to us again and there is no mine or gold," added one of the women.

"Well, there is gold there," I answered. "We have proven reserves that will last a while."

"Well, that's a relief. Thanks, Mac." There seemed to be a collective sigh throughout the room.

"If that's all the questions, then turn on your phones and get cracking," Bert said.

"Bye, folks," Roger said. As we turned to leave, the phones started to ring.

"Come back here on Thursday. We'll have some more money for you," Bert promised.

"Sounds good to me. Roger, lets go to the hotel and have a drink."

"Roger, what was that all about? What's in the back room that I'm not seeing?" I asked, taking a large swallow of Chivas Regal.

"This is one of Harold's boiler rooms. They hire these people to call the mailing lists and return any calls from the advertising," he said.

"Where are the calls coming from?"

"They advertise across the country. From what we saw today, the ads seem to be working."

"How many of these boiler rooms are they operating?"

"Let's see now, one in Portland, Oregon, one in Kansas City, another in Chicago. This one and, of course, the office in Beverly Hills. That's five and I heard they are setting up something for Houston."

"Jesus, Roger. How much are they selling?"

"I don't know, a lot I guess. Maybe I should find out."

"Maybe you should," I suggested. "Meanwhile, I need another drink."

"Order one for me, too. I'm going to call a couple of girls I know. Back in a few."

We arrived back at the sales office about eleven o'clock on Thursday. Through the glass entry door we could see Danny waving his arms at three fancy suits. "Let's not go in, Mac. We better wait out here," Roger cautioned.

"Okay, but what's going on?"

"I'm not sure, but I have a hunch it's not good."

Danny caught sight of us and pointed while he was talking to the business suits. One of them, the biggest of the three suits, turned and nodding to Danny, came towards the door.

"Maybe we should run," Roger offered.

"What for? Let's see what this is about."

The big guy came through the door and said, "I'm working for the Georgia State Attorney General and would like to have a few words with you. My name is Sam Woods." He handed us his card.

"Okay," I gulped. "Let's have a cup of coffee."

"What's going on Mr. Woods?" asked Roger. We were settled in a nearby coffee shop.

"Our office received a request from Chicago asking us to check out the operation over there," he said gesturing towards Danny's building.

"How can we help you?" I asked.

"Do you know Danny?"

"I met him once in Los Angeles."

"Do you know who he is?"

"He's Harold's brother is all I know."

"Is Harold in Los Angeles?"

"I guess so. We saw him just before we left to come out here."

"Have you talked to Samuelson or seen him?"

"I never heard of Samuelson."

Woods turned to Roger, "Have you seen Samuelson?"

"No, I have not," stated Roger.

"You have my card, Mac. Will you please give me a call if you can think of something that might help?"

"Okay, but I don't think I can be of much help because I really do not know what you are looking for."

"Well, a word to the wise, just keep your eyes open. Don't turn your back on this group."

As he went out the door, I said to Roger, "Damn! Looks like Danny has a problem brewing."

"Maybe, so! Let's go get our money and blow this Georgia place," Roger suggested.

We were on a non-stop flight, Atlanta to LAX. Danny did not have any money for us. He said they would deposit it in our account as soon as it cleared from the sales they had made this week.

"Roger who was that guy, Bert?"

"Bert was Harold's partner. I knew him when I worked for their other company in Hollywood. That was the famous Mr. Samuelson. You were not supposed to know he was involved." It was obvious that I was dealing with Goldstein and Samuelson. Several years ago they had been indicted and sentenced for selling unregistered securities. I woke up about 35,000 feet over Texas and thought, what else am I not supposed to know?

"How was the trip to Atlanta?" Mary asked over the phone.

"Interesting. A little different than what I expected. The Georgia Attorney General showed up and asked me a lot of questions that I couldn't answer."

"Maybe you should go see an attorney?"

"That sounds like a plan. We'll do just that."

Roger came in as I hung up. "Is there an attorney we can get hold of on short notice?"

"I think so, Harold has a firm just down the street. They are on retainer I think."

"Get us an appointment as soon as possible, okay?"

"I will get right on it. Maybe I should go down there now."

Clay's brother, George, had flown to Ciudad Obregon, and picked up the president of the bank. They were driving around part of the ranch. "Well, you've seen most of what I can show you from the ground. Cows look in good shape, don't they?" George said.

"Yes, they do. What are you going to use the loan money for, George?"

"We have a contract with a company in the States that will buy our gold. If you will lend me the money on the herd, I'll put it into the mining operation."

"I think I've seen enough. Let's go back to the bank and draw up the papers." George jerked the steering wheel of the Jeep and just missed hitting a large saguaro cactus.

"Are you okay? You look pale," the banker asked.

"Yeah, I'm okay. Just a little dizzy spell."

Later as they left the house to go down to the plane, George said to his wife, "I haven't been feeling very good. I'm going to stay over tonight and go see the doctor tomorrow."

They were under full power climbing over the sharp rocky ridges on their way back to Ciudad Obregon. "You know, after all these years, we finally have found a way to get these mines in operation."

"You'll put a lot of people to work, too. This will help the local village," the banker said.

The pain kicked George in the middle of the back. The muscle reaction pulled the controls to his chest. The banker had never flown before and even if he had there would be nothing he could do. The 182 angled straight up, hung on the prop for a moment, stalled, and fell into a full power spin.

Roger and I walked down to the offices of the attorneys who had originally given Harold their opinion that his plans for gold sales was legally sound. Their specialty was tax and securities. One of the attorneys removed his horn-rimmed glasses and gave Roger and me a long searching look. "The firm did a lot of research for Harold on the present securities law. In our opinion, the trading that U.S. Gold is presently doing is within the law."

"Can this be considered futures trading?" I asked.

"If it was deemed so, then all they would need is to apply for that license. We don't think it is."

"Apparently they are taking silver bars in trade for the gold. Is this proper?"

"Silver is a commodity," the attorney said. "It is legal tender. If the owner of the bar wants to trade his personal property, sell it, or give it away, he has a right to do so."

"Then there is no wrong-doing on this operation?"

"In our opinion, no. There is no difference here than if the investor bought the gold on the open market. Look at it as a stock purchase. If you buy stock in a company you hope the stock will rise in value. The public in this scenario, is trading their silver expecting a gain on the gold."

"So the bottom line is the delivery of the gold."

"That's correct. There is no legal problem until someone doesn't receive what he was promised."

"Harold then, is borrowing the silver from the customer and giving him a promissory note."

"That's it, the promise is to return him gold in exchange for the silver. If the customer accepts that condition, then Harold has their silver and they have a piece of paper. People dealing with commodities, stocks and securities are considered sophisticated investors. They do their homework, the due diligence studies, weigh the pros and cons and either buy or not. Remember how it is said, let the buyer beware."

"Okay, I guess we're clear."

"Yes, you are but for one minor consideration."

"And that is?"

"If Harold doesn't deliver the gold?"

"Why would he be so stupid as not to deliver?" I wondered.

"Just hope that he does, Mac. I've got another appointment, guys. Get out of here."

"Thanks for your time," Roger added.

Back at the office I had a message to meet Morales at the Playboy Club. The parking on Wilshire Boulevard was a problem, but it was worth it to get to watch the bunnies. "I had a meeting with Roy Burkett yesterday, Alphonse. He's the guy from Arizona that has the pulverizer we need for the refinery." A nice cottontail bounced by our table. I lost concentration on the meeting for a moment, then said, "He has a machine that grinds up garbage and is in need of some financing help. I thought you might know of an investor for something like that."

"I'm working on the environmental waste problems now. I would be interested for my own company. Exactly what does he want to do?"

"He has a contract with Pima County in Arizona and wants to build a bigger machine. I've already talked with Dick Nichols. He's agreed to build the unit in his machine shop."

"Give me Roy's phone number, I'll call him. Would you like another Scotch?"

"Sure, I'd like that bunny over there to serve it."

"How is my airplane working out?" asked Alphonse.

"Mortimer? He's fine. Just had his annual check-up. We have him tied down at Tucson International."

"Glad to hear that," buzzed the snake. Sand and cactus were no match for a playboy club but sometimes the club contained rattlesnakes, too.

It was late in the evening. I had gone back to the office to check the answering machine. Alphonse was leaving by the back door. He was carrying two big bars of silver.

On a fine Sunday afternoon what better thing to do than have a party on the deck of a pretty sailboat in the bay at Newport Beach? We were all relaxed and the beer was quickly disappearing. "How are you doing with the silver thing, Richard?"

"We are having a little trouble finding enough scrap, Mac. Seems as if everybody and his brother have caught on to the fact there is some value in it. Bunker Hunt is trying to control the commodities market and Pacific Coin is gobbling up everything in site for their bagged silver coin business."

"How about the industrial film?"

"Someone is making deals with the hospitals and the industrial areas to take all they have on contract. The x-ray film was a good source, but the doctors are recovering their own now," Richard replied.

"That doesn't sound so good for you."

"It's not. I'm hoping you can get these mines operating soon. We need another source of supply."

"Clay told me about a high grade silver mine near Culiacan, Mexico, when he was here last week. Might be a good idea to arrange a way to go have a look at it," I offered.

"That could be an answer. When can you talk to Clay?"

"Not for a few days, he called last night and said they were going down to his brother's ranch in Mexico. His brother, George, had an accident of some kind," I said.

"That's too bad, hope it wasn't serious."

"Ahoy, request permission to come aboard," boomed a voice from the dock.

"Hey, Roger, permission granted," said the Captain.

"I want you to meet my friends from Hollywood. This is John Flynn and his wife, Christy."

"Hi, John, you guys like a beer?" Then I added jokingly, "You live in Hollywood. Did you ever meet Errol?"

John grinned, "Once or twice. He was always drinking a lot and didn't show up too often. He was my father."

"You're kidding me," said Richard.

"No, it is true. I was pretty young when he died, though. Really do not remember too much about him."

"Must have been interesting living around all the movie stars," Roger said.

"It wasn't any fun, Roger. Most of those people are flakes. They live in a dream world."

"What are you doing now, John?" asked Richard.

"I'm going to school, trying to learn to be a director," he replied.

"My dad produced a movie a few years back," I offered.

"Is that so? What movie?" John asked.

"The title was Bride of the Monster. I think it was also called Bride of the Atom. Ed Wood was the director."

"I know that one. The lead was Bela Lugosi and Tor Johnson played the monster," John replied.

"My brother had the part of the reporter. That's why my dad financed it," I said.

"I heard recently it may be nominated for an award," John said with a sly smile.

"That's interesting. What award?"

"For the worst picture ever made. Ed Wood may be famous, too, someday. He's a transvestite you know," John continued.

"Swell," I mumbled. No wonder my old man never wanted anything more to do with the movie industry. I had met Ed Wood, but never could have guessed that he might be wearing pink panties. "Enough of the movie world," I said, having just had my ego completely punctured.

"I'm getting nervous about being in Hollywood, myself. Looks as if there will be a strike by the Actors Guild soon. I'm looking for a job. Thought maybe I might sell silver and gold. I know a lot of people around here," John said.

"Say John, you've been living here all of your life. Did you ever hear of any problems with Harold Goldstein?" I asked.

"Why hasn't Roger filled you in on those characters? He worked with them 'til the end. They sold securities illegally and were convicted of fraud about five years ago. The company was Goldstein-Samuelson. Harold just got out of jail. He's on probation."

The silence of our little party was deafening. Roger lit a cigarette and gazed at nothing across the bay.

I could sense the weight of the Sign hanging around my neck.

"Damn it, Roger! Why didn't you tell me about what Harold had done?" We were in the office and the atmosphere was the same as being in the center of a very black cloud.

"Ever since I've been in this country, I've gotten myself involved with the wrong people. You guys are the first that I met that were trying to do an honest deal. Harold put me to work with you as a kind of spy. I reported to him on what your plans were. When I realized you were working on real things, I wanted to be on your side. I thought my life would get straight."

"Where is Harold now, Roger?"

"He's in his apartment. After the rumble in Atlanta, he's staying out of sight."

"Let's go over there. I want to talk to him about this bullshit!"

I am a slow thinker. If I'm in a chess game, the correct move finally appears; and if the moves are not forced by a clock, I can generally hold my own. This chess game was different. I had all the men placed correctly on my side of the board, but had never played in a game where you were not allowed to see the opponent's side of the board. The clock was ticking.

"Harold! What in hell is going on! I need to know and I need to know now!"

My pink-cheeked little benefactor put his hands in his pockets and started to pace the room. "Mac, I've been made a victim of the government. The SEC convicted me on a technicality that they had invented to put a halt to the trading I was doing at G & S. The government never proved that we had broken any existing laws. They found me guilty only on the new one that someone came up with to cover my ideas. Even that was questionable. They put G & S out of business. A lot of investors lost most of their money because of the SEC not because of me. Look at me, Mac. I'm only twenty-six years old. How can I be smart enough to cheat the government of the United States?"

"What about this probation thing, Harold?"

"When they let me out of prison as part of the original sentencing, my probation control said I was not to work in the securities business. I'm barred from being a stockbroker for ten years."

"How does that affect this operation?"

"According to the attorneys, it doesn't. We're not dealing with securities and I'm not an officer of any of the companies."

"So you are considered an innocent bystander?"

"Yes. I receive a consulting fee to offer my advice on the buying and selling prices of the metals. There is nothing that the company does that is illegal. We spent a lot of money researching this area. We hired the best advisors we could find. I understand you talked to our attorneys the other day. They are considered to be the best around here. Didn't he tell you we were within the law?"

"Yes, he did. But I am still concerned with what I learned in Atlanta."

"The attorneys suggested that I disappear from the operation here for a while. They think the only reason the government is sniffing around is because of my presence."

"Obviously the government doesn't like you," I said. "What are your plans then?"

"I don't really like this mining thing. It's too slow for me. I need more daily action."

"Development of underground workings takes a while, Harold. The snow is still ten feet deep at the Gold Lake. We won't even get there 'til next June."

"That's what I mean, Mac. This is too slow for me. I've found a frozen food company in Chicago that I can take over. In fact, the owners are coming in tonight in their Lear Jet to take me back to look it over. Would you like to come along?"

"No thanks. I think I have enough to do around here. Right, Roger?"

"Looks like we'll be scrambling for a while," Roger said. He really did want legitimate work.

As we were leaving, Harold said, "Say, Mac, lend me a hundred will you? I'm a little short."

Sometime later I learned that the chauffeur driven limousine the executives had used to collect Harold, and the company Lear Jet had been rentals for the day. I learned this because Gold Lake Mining Corporation got the bill. The promoters had promoted the promoter! I never got my hundred back.

"Alphonse, what do you think of these developments?" I asked.

"The government seems to have a 'hard on' about Harold," he answered.

"What is your position in all of this?" I asked. I was a few weeks late in doing my due diligence studies.

"I have no problem with it. I was hired by Harold as a consultant. With the questions being asked by the officials, I've elected to resign my position. I have no conflict with Harold. I can work for Gold Lake. Do you want a consultant, Mac?"

How can it be that everyone is a consultant? Somebody has to be in charge of something. Are there no bosses?

"Let me think about that for a while, Alphonse. Where do we stand in this business right now? What would you recommend that we do?"

"There is no change. The sales are going well and you have the mines almost prepared for production. Just keep going. You don't have a thing to do with Harold's problem."

"We'll still need more money for wages and supplies. Where does that come from?"

"I can watch that part. By the way, I heard that Dick took the plane to Mexico. I don't want him flying my airplane. I want you to get it and park it some place where he doesn't have a chance to go joy riding."

"We're planning a trip to Mexico to look at a high grade silver mine soon. We'll take Mortimer then," I answered.

"I might be in Arizona, too. Maybe you can introduce me to Burkett. I want to see his garbage machine."

Our whole crew was busy working on each project. There was a feeling of competition among the guys. Each was trying to be the first to get their project into production. The key was the cash flow. If the money was available, the mines could produce. The prices of both gold and silver were on the rise. So was inflation.

The journey to Mexico was set. Richard and I were at Mary's place outlining our options on the development of the silver source in Mexico. Clay drove into the yard. Coming through the door, he said, "That damn Dick is the shits."

"What's wrong now, Clay?"

"He won't give me the key to the plane. He says, wherever the plane goes he goes. He is demanding that he come on this trip, but can't leave 'til next week."

"He's in control of our business with that attitude," Mary said.

"Is Harvey ready to go?" I asked.

"Yeah, he rearranged his schedule to go tomorrow."

"Anybody have any ideas?" I asked.

"Call Triple A and have them make a key," Mary offered. "I have a card."

At four a.m. the lock and key man parked next to Mortimer. He extracted the door lock, found a number and made a key. The same key fit the ignition. The whole operation took twenty minutes. We gave the man twenty dollars. He said, "Thanks, call me anytime," and drove away. He never asked us for identification. He had no way of knowing if this was our plane or not.

"We're in the wrong business, Mac. We should start stealing airplanes," Richard said.

We left without Dick.

"There it is. See, along the river bank near the church," Clay said. He had been here before. We were high over Copper Canyon of Mexico. The mesas on each side of the canyon were not unlike our Grand Canyon in Arizona, except for the thick foliage. This Canyon is in some ways more beautiful. The layered walls were alive with the vivid colors of the copper minerals. We could see portions of the railroad bed that had been chiseled and blasted out of the solid rock.

"There on the north side is the famous gold mine, the Lluvia de Oro," Clay pointed out.

"That means rain of gold, Richard." I offered.

"The mine was fabulous. The wire gold was so concentrated, when they blasted, the rock wouldn't fall. It hung from the backs of the drifts and the miners used axes to cut it loose," Clay explained. "They have been working here for a hundred years."

"Where's the silver mine, Clay?" I asked.

"Over there to the south. It's interesting that here the gold is found on the north side of the canyon and the high grade silver on the south."

Far to the west we could just make out the Pacific Coast line and the city of Mazatlan. We landed on a dirt strip near a small village.

"Welcome to Choix," Clay said.

"This doesn't look too choice to me," offered Harvey as he cut the switches and the prop wound down. We were being approached by some mean looking guys in olive uniforms. They were all carrying automatic rifles.

"This is the Mexican army. Wonder what's going on?" Clay said.

In rapid fire Mexican, the Captain ordered us to stand in line while they checked our papers. Two of his men opened the cargo doors and crawled inside Mortimer with their weapons. Clay was the spokesman for us. He understood the dialect down here. There was a marked difference from the border lingo.

We had brought four cases of powdered milk to give to the village general store at our final destination. When you go into the remote mining areas, it is protocol to bring the local people something that they need and is hard to get. Clay had learned milk was needed for the kids. In our world a priority was a modern newspaper. These gestures opened doors in worlds where most of the time strangers were not welcome.

Everything was found to be in order after Clay gave the Captain fifty bucks. The guerillas wandered away and squatted in the shade.

"Why the army, Clay? What did Pancho Villa say?"

"El Capitan says they have a joint exercise under way with the US to close down the marijuana and poppy growers' operations."

"Where are these growers located?"

"All around this area, but mostly where we are going to the mine. El Capitan suggested that we be very careful not to leave the trails."

"Super!" says Harvey.

"What about the milk?" Richard asked.

"He didn't care, since it's for the kids."

Probably the fifty did not hurt either. We had landed here first so Richard and Clay could arrange for sleeping quarters and reduce our payload. The flight into the canyon to unload would be a quick twenty minutes. The landing field was short and the smart thing to do was unload some of our weight. With all of our tools and the milk for the kids we had too much weight to land on the dangerous strip along the river. We would be way overweight to land there with an extra four hundred pounds of people.

The elevation difference from the canyon rim to the river was about fourteen hundred feet. There was enough room between the walls that we could circle down to the river. As we prepared for the descent a large Huey helicopter appeared. The crew was checking us out. One soldier was taking our picture. We smiled, said cheese and waved. It was an US Army chopper!

"Guess we passed inspection," Harvey said as they left.

"Yeah, they didn't shoot."

"I'm going down river after we get to about three hundred feet and turn on final near that vertical cliff," said Harvey as we turned between the canyon walls.

"What's that over there, Harvey?"

That 'over there' was part of a wing that used to belong to an airplane. "Guess we better pay attention to this approach, huh?"

"Suppose so. What approach speed do we try for?"

"I think about sixty five. We'd better use full flaps." Normally Mortimer set down between eighty-five and ninety miles per hour.

This strip had been built on the river's edge about ten feet above the water level. Four to five hundred feet down the track there was a bend to the right. The other end of the strip was not visible. "Give me flaps as I call for them, Mac. As soon as we are down, get 'em up. We don't need to throw rocks through them."

"Aye, aye, Capitan."

"Help me with the brakes. We have to get slowed down enough to make that turn." The trick on this strip along with holding down the landing speed is to touch down on the very beginning of the landing strip.

The evening is calm, no discernable wind. The turn on final

approach is smooth and old Mortimer is girding his loins for one more rough, rocky, dusty, and dangerous landing. So are his passengers.

It is quiet on approach. The wings dip a little from one side to the other. The concentration on locating the point you want the wheels to hit is intense, while at the same time, scanning the scene for some hazard such as a burro, ox or goat looking for a blade of grass. The touchdown point seems to hang in suspense for a moment. You are actually looking at your future at that spot. Every landing is a controlled crash.

The world of reality is back in an instant. The gear jolting to the ground, the noise of rocks hitting the aluminum skin, the awareness of how close a mountain is to your wing tip, and how fast that mountain is going past, while the nose is settling down enough to allow forward vision. No way to know how long you have been holding your breath while equalizing pressure on each brake pedal, trying to keep your machine moving straight ahead.

That bend ahead is getting here too quick! Slow down, slow down! No way to allow a skid on turning. The river is over there. Watch the wing tip, more right pedal. Slow down, Mort, slow down!

Lots of dust and dirt, but the noise is getting less. We are around the turn and there ahead is a large pile of big rocks. Oh, shit, Mort! Let's stop now, okay? As if he was listening, he complied and slowed to a halt about thirty feet from the end of what was called a runway. Next to the rocks was a crumpled Beech Bonanza.

"He didn't do it right," Harvey said nodding at what was once a nice airplane. Taxiing back the way we had come, Harvey said, "Next time we'll put Mort down closer to the end of the strip. We flared too late."

"That sharp bend is the shits!" I agreed.

"Let's off load as quick as we can. We must get back before dusk," Harvey ordered. The cargo besides the powdered milk was fresh water, backpacks and canteens. The rest of the trip from this point to the mine site would be on foot.

"I'll be damned, look at that." There was a church built out of rock, bell tower and all, setting on a rise just beyond the little store.

An old Mexican appeared with a beat up wooden cart pulled

by a grungy burro. “Buenas tardes, Senores,” he sang. “Bienvenidos.”

As the cart disappeared behind the rocks on its way back to the store, Mort spun his prop, coughed and prepared for the flight back to Choix and our penthouse suite for the night.

At Choix we endured the ritual again. The same Capitan, asking the same questions as his little army rummaged about inside Mortimer. This time no money changed hands.

It was about a half-mile into the village. We walked down a dusty trail that could pass for a road and found Richard and Clay in a cantina with a liter of tequila on the table in front of them. “So how did you find the conditions in the canyon?” laughed Clay.

“Not the same as the roller-coaster at Pilar, but it’s still exhilarating,” I said.

“Any cold beer?” asked Harvey.

“No, but it still tastes good,” replied Richard.

“I’d like the tequila,” I said as I reached for one of the tin cups. “Learn anymore about the army, Clay?”

“Yeah, the US and Mexico are running a joint effort to shut down the growers. El Jefe tells me they haven’t found any of the people yet, but they are burning the fields.”

“Is it safe to go into the area?” Harvey asked.

“They know who we are now. There will be no problem if we keep to the main trail. The jungle on either side of the trail has been booby trapped with various devices,” Clay offered.

“That doesn’t sound good to me,” Richard said. “Maybe we should take guns with us.”

“No, no guns. The guards will think we’re the army looking for them,” Clay explained. The noise of a gasoline engine start-up filtered into the cantina and the light bulbs over head brightened.

“What’s with the church in the canyon, Clay?”

“The prospector that originally found the big gold mine built it. When he got rich, he did that to repay the Lord for guiding him to the strike. This spot is a famous shrine for the area.”

“Did not look in very good shape,” I offered.

“No, it’s not. It’s doing well for being about sixty years old.”

“I’m off to bed. I’m tired,” Harvey said.

“Good idea. We have to be out of here at dawn,” Clay added.

Our room was nice compared to many other spots I had been in this country. Richard and I were bunking together.

"Hey, we even have a shower and a flush toilet," remarked Richard as he placed the kerosene lamp on a table.

The cockroaches were scurrying to get away from the light. "Better shake the sheets to clean the spiders and bugs out." Walking into the bath, I noticed a large brown shadow on the floor. A closer look revealed a live tarantula. He was quietly waiting for some supper. I had a feeling that upon seeing me, he just blinked his eyes and yawned. He was not concerned at all.

Finished with brushing my teeth, I said to Richard, "When you go in there, don't step on our new friend."

He peeked in the door, "Bullshit, I'm not going in there at all!"

"That guy won't bother us. He only wants something to eat."

"Maybe. But I don't want to give him the chance to take a bite of me."

"He's no problem," I replied, turned over and went to sleep.

The sky was trying to brighten some in the east. I had followed my own advice to Richard and shaken out the sheets, but the fleas had not left with the spiders. My legs below the knees had served as lunch for the little guys. Richard was sitting up leaning against the wall, staring at the washroom area. "Sleep okay?" I asked.

"Not at all. I've been watching for the tarantula."

I went over and looked in. The spider was rolled up into a furry ball. He had had a good night's sleep.

"Let's see if we can improve our performance from yesterday," Harvey suggested. We were all set up with the proper pitch to the prop and flaps extended and rapidly approaching the end of the strip. The church floated by to the south. "This time we'll power straight in. Let's put the nose wheel in that bush on the very edge of the strip and give me three quarter flaps." We hit with a thud and the noise again sounded like an avalanche falling on a tin roof. Mortimer had been here before and had learned how to handle the dogleg in the middle of the course. This landing seemed simple compared to yesterday.

The walk to the mine was not a Sunday stroll in the park. One of the muchachos who lived at the shrine had offered to help us find the mine, for a fee, of course. The trail followed down river about three miles then turned south into an arroyo before heading into the jungle. The humidity was high and there was no breeze blowing. The sweat rolled off us. "Who said this was going to be a romantic adventure?" I queried. No one answered. The trail was steep and the loose rock rolling beneath each boot, made every step more difficult.

Instead of having a quiet walk in the woods, we were being bombarded with an argument about our presence from thousands of parrots. The noise was almost unbearable. There was a prickly feeling that we were being constantly watched.

After three hours of this torture, I begin to notice a change in the color of the rocks in the dry streambed. Venturing off the trail about ten yards, I retrieved a reddish rock with a gray coating on it. Checking with the magnifying glass, I could see the coating was native silver.

"Richard, take a look at this," I said. He had stopped to puff while I was messing around with the rocks in the arroyo. "What do you see? This gray coating is pure silver. I never saw anything like it before!"

"Most of the rocks in the creek are coated like that, Mac. Look up the creek, There is more of it."

"You're right! We must be getting close to the mine. This is beginning to look like rock from the dump."

From this point on, we plodded in a dream. The mountain grew steeper and the heat became more oppressive. Every step was a huge effort. The silver lying on the ground was unimportant. I never even took a sample of that rock. The priority was just a spot to lie down in the shade and cool off. Almost every hour someone had asked Clay how much farther. Each time the answer had been, not far, now. How far is not far, I wondered.

We had been on this trail for six hours. We finally arrived. Standing in a line in front of the mine's cabin were four big guys with rifles.

"Clay, what about this? You said no one was working the mine. Who are these guys?"

"I don't know. Better ask though before we get shot."

Clay talked with them awhile then waved us on. "They have some cold beer and frijoles for us. We can spend the night."

"What's the deal?" I asked.

"I think they are running the poppy farms and use the mine as a cover for the operations. Be careful what you say. Don't ask any questions." Every one of us was so tired from the trek that it did not sink in right away that we were in the right place at the wrong time.

The cold beer was a real treasure. Harvey, Richard and I retreated to a flat rock and sat down well out of earshot of the motley crew we had just met.

"Appears as if we may have a problem," I said.

"Clay assured me no one had been working here for several years," Richard groaned.

"I doubt these guys care about silver mines," Harvey said. "They don't look much like miners."

"Probably they make a few more dollars growing the crop than digging up rock," I said.

"Damn, I'm tired. Think they have another beer?"

Clay came over to where we were sitting, "I want to show you something you've probably never seen. Watch." He hammered and hit the rock he had brought. Instead of breaking, it just flattened out.

"Let's see that rock, Clay," I said reaching for it. The damn rock was at least fifty percent wire silver.

"There is a problem with this ore," said Clay. "It has so much metal that you can't crush it in a jaw crusher for processing."

"We need a problem like that," said Richard.

"When can we see the mine, Clay? I want to see how wide the vein is."

"We ain't seeing no mine, fellas. These guys want us out of here. We leave at dawn."

The silence was deafening.

The sun was up about twenty minutes. The morning air was about fifty degrees cooler as we started down the mountain. "Richard, did you notice what we were sleeping on?"

"No, it was dark in that room. I didn't notice."

"Those were cases of Dupont gel forty percent sticks of dynamite," I told him.

"Jesus," he said. "The tarantula isn't so bad after all!" Everything is relative I suppose. The trip down was easier, only a four-hour walk.

Back at the church, we were offered plates of frijoles and tortillas. "Would you like some of our local drink?" El Jefe asked.

I took one whiff of the concoction and quickly begged off. Richard tried a cup full. Turned out to be fermented guavas. Richard was sick for a week after that cup full. Sometimes discretion is the better part of valor.

The flight back to Arizona was uneventful. The adventure had produced nothing but blisters on eight feet and a terribly upset stomach that Richard was painfully carrying around. Clay got the Sign this time. He could have learned about the narcotics operation had he asked.

About this same time a discussion was taking place in Chicago. "I think we have to learn some more about this U.S. Silver operation," the Assistant District Attorney said to his aide. "Issue subpoenas to those people on the coast. We'll question them about Goldstein's operation."

"All right. I'll get the motion to the judge by tomorrow." A small wheel of justice had started to grind.

"There is a national movement on the environment, now. The cities are being pressed to clean up the garbage dumps," Roy Burkett was explaining to Alphonse. The three of us were meeting in Roy's apartment in Tucson. He was pitching to Alphonse, trying to get backing for his solid waste project.

"I'm very interested in helping with ideas to clean up our environment," said Alphonse.

"The invention will reduce the volume of the trash being dumped in the landfills six to one," Roy said.

"This is important," answered Alphonse. "The new laws are

addressing that very problem."

"I would like to build another machine based on my new patent. The new one will do a better job than the prototype I'm using now."

"I can get the money together to help you. This is an exciting program for me, Roy."

"I'll call my attorney and we can start drawing up the deal right away," Roy suggested.

"No, no. We don't need to spend money on attorneys. I can write up an agreement for this," Alphonse said.

"Well, all right, let's get started."

"Listen you guys, I don't have time to work on garbage. How do I fit in here?" I asked.

"I'll let you have the small machine that you saw. You can use that for the mining operation, can't you?" Roy offered.

"You focus on the mines, Mac. Roy and I will work on the garbage grinding," said Alphonse.

"Sounds fair. I'll go by and look it over."

"Thanks for your help, Mac. See you later," said Roy.

On the way back to southern California, Richard and I stopped by Lukeville, Arizona. Lukeville is a sleepy border town and is a gateway to the Altar Desert just south of Ajo on the edge of the Saguaro National Monument. The Sierra Pinta mine was about thirty miles south near the Sea of Cortez.

"We aren't having any problems, Mac. The rest of the machinery is about ready to run. We plan to start to work on one of the dumps in about two weeks," Robert Montgomey explained. Robert was my Mexican partner and had been watching over the Sierra Pinta operation while I coordinated the other projects.

"Since everything is in order, we won't take the time to drive down there," I said.

"Yeah, no need. I have it covered. Just send money."

"There will be another deposit made for operating funds. We'll send some more cash soon," I promised.

Roger had his feet up on the desk staring out the window.

"How goes the war, Roger?"

"Quiet on the western front. How was the trip?" he asked while taking a drag on the Pall Mall.

"Not good! We couldn't get near the mine. The narcs had the whole state covered. How are Jim and Monty getting along with the refining of Elodia's gold bar?"

"Okay. Jim told me they should be ready to sell it by next week. Slagle ran an ad for help on the boat in Alaska. He has been interviewing quite a few people. Some of them want to sign up for no pay just room and board. They think it will be a romantic adventure."

"Where is Jim?"

"He went up to Susanville to help start up the mill. They need some more money."

"Did the deposit show up from Atlanta yet?"

"No, it hasn't. Danny had another cockamamie excuse."

"Ah shit! Everybody needs some of that cash. We're going to run into trouble if they don't get it here soon." A knock at the open door announced two nice looking people.

"Hello, we're looking for the Gold Lake Mining offices," the fellow said. The girl with him had a great smile.

"This is Gold Lake. How can we help you?"

"We're from the Wall Street Journal and would like to set up an interview," said the great smile. "We're working on an article dealing with gold and silver investments and need some input from someone in the actual mining of the precious metals."

"We don't talk to reporters," blurted Roger.

"Whoa, wait a minute, Roger. Let's see what they are doing. Maybe they can be of help to us," I said.

"You talk then. I don't think it's a very good idea, though," he mumbled as he went out the door.

I apologized for Roger and spent about thirty minutes explaining how the mines worked into what our understanding of the sales and investment program was. The last question they had did not make sense for an article on precious metal mining. The smile had asked, "How does Harold Goldstein fit into this scheme?"

Roger had been right. Don't talk to reporters!

I received another Sign to add to the growing pile.

"What are we drinking, Roger?" I had located him in his favorite watering hole down the street.

"Lots of scotch," he replied. "I'm worried about those reporters showing up, Mac."

"I couldn't see any harm in talking with them until they asked about Harold."

"If they asked about him, then we are in trouble."

"Why? They seemed like nice kids."

"Mac, they are probably investigative reporters. They aren't interested in learning about mining. The only interest they have is learning about what Goldstein is up to. If they write a story, it won't have much to do with what you talked to them about."

"Man, I hope you're wrong. I'm getting so I don't like this modern society. The mountains have more to offer than this political bullshit way of doing things."

"I don't think I'm wrong. I've got a feeling that ain't good about this." His feelings were correct. Two days later we received a summons to appear in U.S. District Court, Chicago, Illinois.

Back in the laboratory Monty had his pots boiling and would periodically stir and offer his incantations over them. "I called this meeting for several reasons," I said. "Let's go over where each of you stand with your projects. Monty, what have you learned of the value of the ore we have?"

"When I extract the metal by precipitating it with aluminum from the hot acid, the values are very high. The concentrates from Alaska are extremely good in gold, but there are higher returns in the platinum group metals," Monty replied.

"What's the next step planned to start a production line?" Roger asked.

"We have the initial heating units and the filters set up. The electrician will be finished with his wiring next week. We are ready for the pulverizer," Monty continued.

"That's in Arizona, now," Monty said, as he poured another boiling flask of hydrochloric acid solution through a filter.

"Bob, we'll run the two barrels of concentrate you brought from Nome as soon as the grinder is in place. How are you coming with your plan?"

"I have compiled a complete list of all the supplies we'll need for the season. As soon as the money is here, I'll start ordering. The shipping schedules are set for the barge from Seattle," Slagle answered.

"When does the boat leave?" asked Jim.

"As soon as the ice starts to break up. Probably the first of May."

"Jim, how are things in Susanville?"

"We made the first run last Friday and recovered some concentrate. Monty is testing that now. I believe that we can get a couple of tons a week. That should give us about fifty ounces of gold and silver," answered Jim.

"Okay, the Mexican mines are progressing, too. Clay was slowed down some because of his brother's death, but he thinks he'll be on schedule."

"Mac, we're going to have to have some more money. All the cash up to date, has been put into machinery. None of us have drawn anything for our personal out-of-pocket expenses," said Jim.

"The electric bills are in again and I have to get more supplies for the lab," added Monty.

"I have some hefty phone bills due and the barge company wants a deposit soon," said Slagle.

"We have been promised another deposit, but it isn't here yet," I said.

Bishop Jim Stokes, the owner of the building, came in the door. "Well, I caught all of you at once."

"Hey there, Bishop. How are you?"

"I'm feeling all right. Heard you talking about money. When can I have some? The payment on the property is due."

"When it rains, it pours, doesn't it?" said Roger.

"We've been assured the check's in the mail," I said. "While you're here, there is a problem that we should address. Your opinion would be welcome."

"I've got a few minutes. Shoot."

"We received a notice from the US Attorney in Chicago to appear at some sort of hearing. It relates to Harold's operations. Roger and I aren't sure if we should go."

"Is it a court order?" asked the Bishop frowning.

"No. The letter states the hearing is for a sales organization and it reads more like an invitation for Gold Lake Mining to help in the discussion."

"Your Company isn't associated with the sales group, is it?"

"No. We've never heard of the sales organization," said Roger.

"Have you talked to an attorney?" the Bishop went on.

"Not about this letter, no. The attorneys that help set up U.S. Gold told us, in their opinion, we had no connection with anyone else," I answered.

"We certainly do not believe we are doing anything wrong."

"Maybe you should go to Chicago and see what it is about, and check it out with those attorneys again," said Jim.

"They must have you connected in some manner. I think it would be better to go over there and listen to what's going on. If you don't, they might find a reason to pull you farther into whatever investigation they're doing," offered the Bishop.

"I think so, too," said Monty. "If they have any questions about what we're doing, they can come out here and look at what we have."

"I'm on the run, guys. Get me some money." The Bishop waved as he walked out the door.

"When's the hearing, Mac?" Slagle asked.

"Next Tuesday, ten a.m."

"Let's get back to work," said Jim.

"You should have an attorney present for this, Mac. I'll go with you." I had gone to see the legal beagles as the boys had suggested.

"We don't have any money in the account," I said.

"That's not good for us, is it?" he laughed. "I could write you up a letter that will allow you to represent the Company."

"That would be cheaper. Give me the letter," I said.

I got another Sign. Rule number one in the legal world: never try to be your own lawyer.

"Hi, sweetheart. How's the weather in Arizona?" Mary was on the other end of the long distance call.

"Still warm," she answered. "What's happening with your big deal?"

"The government sent a request for us to attend some kind of hearing in Chicago."

"Oh, oh. So they are after Harold?"

"I don't know, but every one over here thinks that I should go."

"This sounds bad to me, Mac. I wouldn't go. Let someone else appear. Remember I told you not to get involved in this deal in the first place."

"Yeah, I remember. What will happen if I don't show up?"

"Probably less will happen if you stay home," Mary countered.

"I've never seen how these hearings are handled. I think it would be good education."

"We don't need that kind of education, Fart Blossom."

This girl is so endearing, "There is one more thing I wanted to ask."

"What is it?"

"They haven't sent us our money yet. Could you use your credit card and get a prepaid ticket for me?" A long sigh filtered over the line.

The comfort level on a DC8 was somewhat better than the flights we made with Mortimer. I leaned back with a scotch and relaxed. This should be an interesting trip.

I found the federal building in downtown Chicago, and located the courtroom on the fifth floor. On one side in front of the Judge's bench was a long table with three people and a couple of briefcases. At the other table about eight guys were conversing. On the table were several boxes of records. These were SEC attorneys. It was not what I had expected. This was not going to be a friendly little get together.

"Would the court please rise!" a large uniform requested. The lady with her recording mechanism sitting in front of the bench began to type. "This hearing has been called for by the Securities and Exchange Commission. The defendants are U.S. Gold Sales, Chicago; U.S. Gold Exchange, Portland, Oregon; U.S. Gold, Atlanta, Georgia; Gold Lake Mining, Los Angeles."

"Is everyone present?" the Judge asked.

"Excuse me, your honor," I stammered. "I'm here for Gold Lake." This was not where I belonged.

"And who are you?" asked the Judge.

"My name is Eugene MacDonald. I'm President of Gold Lake Mining."

"Where is your attorney?"

"We have no counsel present, your honor. I have a letter here requesting permission to represent ourselves." This situation was out of my league already and I had not been here ten minutes.

"Let me see the letter, Mr. MacDonald." I moved toward the bench and nervously handed him the document prepared by the attorney. "This is not proper procedure, Mr. MacDonald, but since you have come so far, I am inclined to grant your request," he said peering over his glasses. I nodded. "However, I strongly suggest that you retain counsel."

"We have no funds to hire anyone," I answered.

"Very well. We shall proceed. Let the record show that Eugene MacDonald is representing Gold Lake Mining Corporation," he directed as he eyed me sadly.

The morning passed in a fog. I did not understand the procedures and most of the legalese was Greek to me. I did not belong here and had absolutely no one to turn to for help. So I focused on the correct way they were addressing the bench.

A witness named Paul Levine was called to the stand. He testified he was the one in charge of all of the sales offices around the US and based in Chicago. The questions asked by the SEC attorneys were specifically directed toward determining if Harold Goldstein was in any way connected to their operations. Levine was not answering any of the questions directly. In my opinion he came across as a professional high class bullshit artist.

The Judge said, "We'll take an hour and a half for lunch."

"All rise," said the big uniform.

In the cafeteria I spoke with Levine. "Say, Paul, tell me what they are digging for. So far all I can see is a witch hunt."

"You're right. They are trying to find a direct line to Harold. The SEC is after him. He is still on probation. If they prove that he is behind this program, they'll put him back in jail."

"This is all they want?" I queried

"Up to this point. The sales and trading procedures that we are offering are all legal."

"Can they shut you down?"

"The government can do anything they want to," he said, "but there are no regulations to interfere with the sales yet. They could invent some though."

"I don't see that our company has any problem here. We have the mines and will be getting some gold inventory soon," I said.

"That's right. Our attorney is surprised that Gold Lake was named in the first place."

"If they stop the sales, then we won't get the funds to finish our production capability," I told him.

"I don't think they will, Mac. Let's see what happens this afternoon."

"Thanks, Paul. See you in court."

The afternoon session was moving slowly with what seemed to me to be innocuous questioning by the government attorneys. One of the sales people answered in a way that was on the edge of implicating our company. He said that I had known about a bill that was presented to our company from Samuelson. I interrupted saying, "Excuse me, your honor, but what was said is not correct." I could have kept my mouth shut.

Immediately the SEC had an opening. One of their attorneys said, "Your honor, if Mr. MacDonald could elaborate on that we would like to hear."

"Mr. MacDonald is acting as counsel for his company. He is not required to testify here," the Judge said.

"I would like to clarify this, your honor."

"Let the record show, Mr. MacDonald will give testimony."

The lady with the machine dutifully placed this in the record. I should have had a stiff kick in the ass.

Sitting in the box was a lot different from sitting in the audience. I was instant prey for the predators. "The previous testimony stated that Gold Lake Mining Corporation was integrated with U.S. Gold," the opposing counsel said.

"That is not the case. We are a separate company."

"Who are the officers?"

"Myself and Roger Crawford."

He walked over to a cardboard box and retrieved a sheet of paper. "We show here on record that Mr. Crawford was working with Goldstein and Samuelson. If you are a separate company, why is a person who was associated with people in the past who were sent to jail for wrong doing, working with you?" I had no answer. I could only blink and swallow.

"Court will adjourn for ten minutes. Mr. MacDonald, I want to see you in my chambers," Mr. Judge said.

"Yes sir," I answered meekly. They had us by the balls.

"Sit down, Mr. MacDonald," the Judge said.

"Thanks, your honor."

"This morning I suggested that you retain counsel and you deferred. That was a big mistake. Obviously you have never been in court before."

"No, I haven't. I've never had a parking ticket."

"From what I've heard so far in this hearing, you shouldn't even be here. But by your taking the stand, you put your foot in the manure." He filled his pipe from a hand-carved humidor and lit it up. "Mac, you somehow have gotten yourself involved with some heavy hitters. I see a lot of bad people coming through this building every day and in my opinion, you are not one of them." He blew a puff of tobacco smoke toward the ceiling. "Now I'm going to adjourn this hearing. I want you to go home and write out the background of your involvement with these people. From that I'll make a determination on how to handle your situation."

"All right. I can do that. How soon do you want it?"

"Within two weeks. Give me a number where you can be reached. I may have some more questions."

This guy is all right. I really like him. He smokes a pipe. "I appreciate your concern, your Honor. Thanks." I had not just stepped in manure I was up to my arm pits in deep shit! I had earned another Sign.

Mary picked me up at Tucson International. "Damn, it's hot here compared to Chicago," I said, shedding my heavy jacket. Chicago had been trying its best to rain ice.

"Did you get educated," Mary asked.

"I got a crash course in how not to do things."

"I told you so!"

"Why is it you're always right, sweetie pie?"

"I work at it," she replied with a wry grin. "What now?"

"I need to talk to Burkett. We need that grinding machine he has."

"But Roy, you said I could take the little machine when Morales was here," I said.

"If you want the grinder, you'll pay for it."

This old son-of-a-bitch is a crook, too, I thought. "This is not what I understood. So how much do you think its worth?

"I want ten thousand dollars." He did not tell me what it was worth, only what he wanted for it.

"I arranged a deal for you with Morales and Nichols to get your new unit built. You don't think that was worth anything?"

"Any deals I have with them is none of your business. You don't have a written agreement with me."

We needed this grinder for the refinery and soon. "Would you sell a license to use your invention in Mexico?"

"I would be interested in that. I own the Mexican patent rights." His eyes lit up.

"I know a man in Mexico that might want to utilize the idea. If I can get him to buy a license from you, could I take the little machine?"

"Bring him up. I'll talk with him," the old reprobate answered. What is going on? I can not trust anybody.

I had called our man in Mexico and told him we would fly

down soon. Harvey had completed the preflight check on Mortimer the night before so we were able to make our departure just as it was getting light. George's widow and her daughter were on board with us. They had come to Tucson to buy serum to counteract the pink eye in their cattle herd. We had agreed to take them back home with the serum.

"There may be a problem getting this serum through Mexican customs, Harvey. Why don't we go to the ranch first, then check in for our entry papers?" Normally we would stop in Nogales or Hermosillo for proper clearance. "If we unload at the ranch quickly, we might get away with it," Harvey agreed. "Let's try it. We'll save some time and probably a nice payoff to the officials, also."

We did not have much cash and expected to come back the same day. Actually, we were smuggling the vaccine, but it did not affect anyone else and the shots would keep the disease from spreading.

We got across the little bridge at the ranch okay, and unloaded the girls and cargo while letting the prop turn. No one mentioned the accident George had had.

Airborne again I asked Harvey, "Shall we go over to Hermosillo for our papers?"

"Why don't we get clearance at Culiacan. It's much closer and we save a lot of fuel."

"Is Culiacan a point of entry?" I asked.

"Yeah, well, I think so."

"Okay, let's do that. We can get back early this evening." We leaned back, poured the coffee and droned our way south to the Culiacan airport.

The airport at Culiacan is flat. The visibility today was unlimited and no other aircraft in the air were to be seen—a serene sight in the desert. We had a straight approach from the north and touched down on smooth concrete that stretched ahead of us for about a mile. "Think Mort can hack this landing field?" quipped Harvey.

"I think he is insulted. Maybe we should throw a few rocks around to make him feel at home. Mortimer slowed and taxied to the fuel dock near the terminal. "Pretty quiet. I'll go see if I can scare up someone."

The first person I located was asleep in the fuel truck. "Senor, would you fill the tanks on the Cessna? We need about 50 gallons." I asked in Spanish.

"Si Senor, uno momento," he said, stretching his arms toward the ceiling.

In the terminal, I saw four officials in uniform talking in a back office. They were not interested and took no notice of me. I found a phone and called Enrique Monasterio the fellow we had come to see about buying Burkett's license.

Enrique was a brother of my attorney in Mexico City who had set up the Mexican Corporation to do the work at the Sierra Pinta Mine. He was well known here and owned a large wheat farm. The President of Mexico had picked him to be head of the National Wheat Growers Association.

Over the years, Enrique had developed a strain of wheat that was now producing fifty percent more bushels than two years before. He was the foremost expert in wheat in all of Mexico. He had answered my call on the second ring and said he would be free to collect us in about an hour.

"Did you find our man?" asked Harvey.

"Yep! He's delighted we're here this early. Will come and get us shortly."

"The Mexican is refueling Mortimer. I checked the oil and all."

"While we're waiting, let's get our paper work out of the way." Up to now no one had seemed to pay any attention to us. We could have just left the terminal. But wanting to follow the rules, we walked up to the desk and asked a clerk where we could find the entry forms for the aircraft. I had a technical visa that allowed multiple entries, but Harvey had to get a visitor's visa.

"You have no papers?" the Mexican asked in Spanish.

"No we don't. We came direct from Nogales and want to get clearance here," Harvey said.

"Oooh, Senores," he dragged out slowly. "I will ask El Jefe about this. Wait here, por favor." He went to the back office. Through the glass door we could see a uniformed officer turn his head and stare at us. He was a little guy with an Errol Flynn mustache and was smiling like a wolf as he came toward us. It was too late to run!

"Buenos dias," I said. "We would like to file our entry papers for the aircraft, and this man needs a visitor's visa."

"Do you have papers, senor?" he asked looking up at me.

"Si, yo tengo. Aqui." I handed him my visa.

"Hokay. These are fine. But I must inform you that this airport is not a designated point of entry. The airplane and this man are here illegally." He turned and motioned to the other uniforms in the back office and said, "Come with me, please." We begin walking toward the ramp where Mort was being refueled. Three other officers in their gold braided caps joined El Jefe.

"Damn, Harvey. Looks like we might have a little trouble brewing."

"Wonder why all the big brass are here." We stood to the side a little and watched while the big boys looked the plane over. They seemed to be more interested in Mortimer than us. El Jefe motioned over to the cargo door side of the fuselage. One of them moved to the left side and opened the single door.

"Maybe they are working on that drug operation we saw up at Choix," I suggested.

The three officers were talking quietly amongst themselves as the other officer came around the prop. He had his hand out and a big smile on his dark face. In the palm of that hand we saw several small kernels. El Jefe said, "Senores. You are under arrest. We are confiscating the plane!" The hand held small marijuana seeds. The fourth officer had planted them. They wanted Mortimer!

"Oh, fuck!" mumbled Harvey.

"Double fuck!" I offered. Where was Enrique? We were going to need some help and quick.

"Harvey, did you see what that guy had in his hand?" They had put us in a little room with only three chairs and a small table. There, of course, was no air-conditioning and it was beginning to get hot.

"I did. There was some lint mixed up with those seeds."

"He had that damn stuff in his pocket," I said.

"They are setting us up. I think they want a turbo-charged Cessna."

"I can imagine what Alphonse will say if we lose his airplane," I mused.

"I ain't worried about the plane. What are they going to do with us?"

"If they want to, they can connect us to the DEA operation in Choix. We have a chance to get thrown in the bastille." The door opened and El Jefe brought Enrique in.

"Buenos dias, amigos," smiled Enrique.

"Are we glad to see you. Seems like we are in a bit of trouble."

"What happened?" he asked, pulling out a chair.

"We thought this was a point of entry. We came directly here with the intention of getting our papers at this office instead of stopping in Hermosillo."

"These guys are friends of mine. We had a few beers together just last night. I'll talk with El Jefe and get you cleared. Shouldn't be a problem. Don't go away. I'll be back in a few minutes," he laughed as he went out the door.

"It's a good thing we know him. Maybe we can get out of here." Enrique came back through the door after about fifteen minutes. He was wearing a frown.

"This is not a simple matter. They have marijuana seeds they say they found in the plane."

"Those seeds are from some place else. The seeds he has have lint and fibers mixed with them. Probably they are from a carpet or his pocket. You can check the floor of the plane, there is no carpet." The floor of Mortimer was covered with rubber mats not carpet.

"They are known for these tricks," Enrique said.

"How do we argue with them?" asked Harvey.

"If the marijuana was planted as you say, we probably argue with money. If it is your marijuana, you are history. No argument," offered Enrique.

"It sure as hell isn't our marijuana," I responded.

"I think we can get you fellows cleared, but the plane is going to be more difficult."

"We hadn't planned on walking back home," said Harvey.

"I'll talk to them some more. I see now that they only want the plane, not you. You're worth nothing to them." He left us alone again.

"That's swell. An airplane is more valuable than we are," said Harvey.

"Give me another cigarette. Wonder if they serve beer for lunch." At that, a big guy we had not seen before entered the little room.

"Hi fellows. Need to ask you a few questions." He was dressed in Levi's and cowboy boots and to our relief was an American.

"Well, maybe we have some answers. First, though, who are you?" I asked.

"I'm working with Interpol," he said smiling. "I got a call that some real bandits were caught smuggling, so I'm checking it out."

"Who are the bandits?" Harvey asked.

"Why, you guys!" he laughed. "Listen, fellas, we know you're not involved with the dope operations down here. We checked you out when you were messing around Choix last month."

"If you know that, why are we being held?" I asked.

"There is one discrepancy in your story that doesn't check. If we can clear that up, I think they will allow you to go about your business."

"That sounds good. What's the problem?" I asked.

"You say that you came directly here to this airport from Nogales. When we checked the time you crossed the border, there is a gap in the time of when you should have arrived here."

"How do you figure that?" asked Harvey.

"The radar picked you up when you entered Mexico and they have a time noted for you. You should have arrived here sooner than you did if you had come direct."

Because of the escalating war on drug smugglers, the US had built an electronic surveillance system along the border from San Diego to Brownsville, Texas, on the Gulf of Mexico. This system was able to monitor any flight. The only way to defeat the radar was to fly very low, up the canyons and behind the mountains. We had made the crossing at normal cruising altitude and were noted on the computers.

"We can explain the time difference," said Harvey.

"Explain. You must have gone somewhere else first."

"We took George Coughenour's wife and daughter home. They were in Tucson and asked us to drop them off," I answered. There did not seem to be any need to disclose we delivered pink eye medicine, too. That would be more fuel added to the fire.

"Okay," he said making a mental calculation. "That would have taken an extra hour. Too bad about George getting killed. He worked with us for a long while," the agent added.

"He must be a legend down here," I said breathing a little easier.

"I'm satisfied with this. I'll close my report. All you have to do is try to retrieve your Cessna," he said with a grin. It is good to have friends in foreign places.

We were released from our cell and went to search for Enrique. He was out by the plane apparently in heated discussion with El Jefe. Both were talking at the same time, waving hands and arms as Mexicans like to do.

It was hot now. Not much breeze and the little bugs were in our ears and zipping around our eyebrows. Harvey and I found some shade and waited to see what the judge and jury would decide. Mortimer was parked at a right angle to the runway. "If we could get to the plane, I think we could take off straight across the field," Harvey said.

"That might be our last resort. Let's see what Enrique can arrange." The pair began walking toward us and Enrique waved us over.

"It's siesta time," he said. "I've invited everyone for dinner. Come with me. We'll go downtown."

"How are we doing, so far?" I asked.

"They are backing off a little on the dope charge, but they think they can confiscate the airplane."

We rode into town in Enrique's Bronco. The officials followed us to the restaurant. A large round table had been prepared and we all sat down and ordered as one big happy family. "Cerveza por todos," El Jefe said hanging his braided cap on the back of a chair.

This scene could happen only in Mexico, I thought. Here we were having cerveza and frijoles with our captors just like one of the family.

"Well, Enrique, what happens now?"

"All of our amigos are going home for siesta, then they want to see us at the airport by four o'clock," said Enrique. "El Jefe is trying to locate his boss in Hermosillo. They want your plane. They have one like it, but it is not a turbo. They like the extra power."

"What do they want, a trade?" I asked.

"They aren't thinking trade, Mac." We were heading north away from the city. "Where are we going? You don't live out here, do you, Enrique?" I questioned.

"I have a friend who might help out. We'll see if he is home. It's not far to his place."

We were driving through low rolling hills on a recently graded dirt road. In the distance were green irrigated fields.

"What do they grow over there? That's not a wheat crop," I asked.

"Tomatoes. Most of the crop goes to Los Angeles. The soil in this area is very rich. One day this estado will be one of the largest agricultural producers in Mexico."

There were no houses or shacks along the road we were following. About a mile ahead we could see a large white building that appeared to be a castle.

"Man, that's a big house," said Harvey.

"This hacienda belongs to my friend," Enrique said. "We'll be very lucky if he is home. He travels a lot."

Large wrought iron gates broke the white adobe wall. As we coasted to a stop, we saw the gates were closed. A large sombrero was crouched in the shade close to the wall. The sombrero came to Enrique's window. Enrique asked if El Senor was available. The guard had a walkie-talkie. He called up to the castle and then opened the gates. Inside the walls was one of the most beautiful gardens I had seen anywhere. This place was an oasis deluxe.

"What does this guy do in the business world?" I asked. "This place cost a fortune."

"Don't ask, Mac. We need his help," offered Enrique.

We went into a spacious room with Spanish tile floors that gleamed with fresh wax. It was very cool and quiet.

"Amigo, mio. What brings you out here to the slums?" a big fellow with bushy white hair said, embracing Enrique.

"These are friends of mine from the States," Enrique explained. "They are having a little trouble with the customs at the airport.

"Oh, what kind of trouble?"

"They made a mistake and landed here without papers for the plane. Our friend at the airport is calling the big boss in Hermosilla. I believe they will come down here."

"How can I help, Enrique?"

"I think I can handle the local officials, but if Hermosillo gets involved, I'm afraid the boys here will lose their plane."

"How well do you know these Americans," he asked nodding toward Harvey and me.

"Mac, here, is a partner of my brother, Eduardo, in Mexico City. They have a mining corporation."

"Why are they here in Sinaloa?" he questioned.

"They have a machine I am interested in for the wheat. They came to explain to me how it works."

The white hair lit a cigarette and pondered a moment. The quiet was deafening. The boss reached a personal decision. "Okay. I'll trust you Enrique. Maybe I can help. Follow me in your truck. We'll drive over to the shop." Outside a light colored Mercedes sedan was waiting. The boss got in beside the driver and said to us, "Forget where you have been, okay."

"We already forgot," I answered.

We followed in a heavy cloud of dust. The Mercedes made a turn off the main road and was leading us over a narrow rough trail through the desert. We drove for at least thirty minutes and then as we came around a barren knoll another airport appeared. The landing strip was paved with asphalt. Parked along the strip were several planes. Four hangers were visible and a new Gulf Stream jet was inside one of them.

"Have you been here before, Enrique?" I asked not really believing what we saw before us.

"No, I haven't. Only heard rumors."

We pulled up behind the Mercedes and got out on the tarmac.

The boss came over and said, "The boys from Hermosillo are on the way. I'm going to try and get them on the radio and see if I can head them off. You wait here."

We did not venture too far from the vehicles. There was this feeling of being watched. Harvey motioned me up to the front of the Mercedes "Look in that second hanger, Mac." Inside was a Queen Air. Four workers were busy around the plane.

"Man. There's a lot of money sitting in there," I said.

"Those guys are masking it off. They are getting ready to paint it," Harvey said.

There was a twin Apache not too far from us. "That twin has a new paint job, too," said Enrique.

"The 207 down the ramp doesn't have any numbers," said Harvey.

"This is a top level operation. What do you think it's all about?"

"I bet they are all stolen aircraft," said Harvey. "I've heard some scuttlebutt around the fields in Arizona about these places."

"Harvey, remember that plane we saw up north where we landed and found the fuel?"

"I bet that plane ended up here," exclaimed Harvey.

"No wonder we're supposed to forget where we've been," I exhaled.

"Make sure you forget," cautioned Enrique, "or someone will make you forget everything permanently."

The chauffeur was standing nearby. Apparently he was left as our guard. The radio in the sedan crackled and we heard in Spanish, "Pedro, send Enrique into the communications room. I need to talk to him." Pedro explained to Enrique where to go.

"Look at that radio tower. Looks to me as if they have enough power to talk to South America," Harvey said.

Pedro moved toward us a little. Under his loose shirt I could make out the slight bulge of a holster. The guy was tough. His eyes were like a cat.

"Damn, Harvey. I wouldn't care to come in here unannounced."

"I doubt that would be too healthy."

"What do you suppose this is all about?"

"I would guess this is part of the marijuana operation. They make a run across the border then change the N-numbers and colors."

"They may be stolen aircraft for sure," I said "and confiscated ones, too. This may be part of the reason they want Mortimer."

In the background there was a rumble of a big generator that supplied electricity for the field. Other than that it was an eerie quiet. This operation was well managed. Enrique came out of the hanger and motioned us toward his Bronco.

I had a major feeling of relief as we left the area. Enrique said, "My friend was able to raise the customs plane. They agreed to meet us at the airport and give you the papers you need."

By the time we arrived back at Culiacan International, it was after four o'clock. The uniforms from Hermosillo had arrived and we were asked to wait. We could see that there was a heated discussion taking place in El Jefe's office. Obviously the local officials were not happy. El Jefe came out of his office and spoke to one of his aides. He looked at Harvey and me with fire in his eyes.

Oh, shit. We are in for more trouble. He stared at us a moment longer, turned and went back to his office. There were to be no interrogations this afternoon. The aide motioned Harvey over to his desk and started to type.

Harvey folded the papers and put them in his pocket. "They gave us clearance to leave in the morning," he said.

"Let's go to my place and have a cerveza," Enrique said. "You can stay with us tonight."

We were sitting on the veranda surrounded by bougainvillea. La Senora had prepared a nice dinner of jalapenos, frijoles and fresh venison. "We messed up your day, Enrique," I said.

"Don't worry about it , Mac. This was interesting for me, too," Enrique said, offering another cerveza.

"What happened in there? Why did they turn us loose?" asked Harvey.

"After we got my friend involved, they had a problem. You knew too much after seeing the operation at the airfield. They could have confiscated the plane because you didn't bring it into the country legally. Luckily they found nothing in the records on you. Since

Mac has his technical visa, the only one they could hold was you, Harvey. If they took the plane and turned you fellows loose, you could have caused them no end of trouble because of your knowledge of their illegal operation. They finally had two options. One, you guys and the plane could disappear. This was difficult because of your knowledge of what was going on with the planes and the dope running operations. Two, they could let you go home. However, there is a condition to this."

"What condition?" I asked taking a large swig of beer.

"You forget everything you saw and heard. If you don't, they will nail you next time you come to Mexico," Enrique counseled.

"They probably will in any event," suggested Harvey.

"Yes, from now on, you will have to be extremely cautious," said Enrique.

"Why did they push so hard? They knew we made an honest mistake," I asked.

"Mac, the government doesn't pay these officials very well. The wages are barely enough for them to buy frijoles. If they are able to confiscate any property, they are paid a percentage of its value."

"If it hadn't been for you, we would have been boiled for dinner, Enrique. Thanks."

"Now, let's talk about the grinder. After all, is that not why you came down here?" he laughed.

"I forgot about the machine. This invention will pulverize most anything. The old man will sell the Mexican rights."

"What could I use it for, Mac?"

"You could mix all the waste straw, corn cobs and the culled tomatoes for soil conditioner to help your wheat grow. If you add the cow, pig and chicken manure, it will make a high class compost."

"Is there one of these machines working now?"

"Yes. The one in Tucson is set up for demonstrating on different materials."

"This idea fits into my plans very well," he said.

"We could take some corncobs back with us and run a demonstration for you if you like."

"I'll get some for you. My wife and I can come to Tucson in a few days. Set up a meeting with your man and we'll talk." We relaxed and slept well. It had been an interesting day.

"Nice morning for flying," I said as Harvey sat down with his coffee.

"I'm going down south to check one of the farms today. I'll drop you at the airport," Enrique said.

"That will be fine. Guess we won't need your assistance today."

We got down from the Bronco, waved as Enrique left and leisurely strolled into the terminal. "I don't feel like a fugitive today," Harvey remarked.

"I feel like we should get out of here," I cautioned.

"I better preflight Mortimer and make sure they haven't done something to it for spite."

"Give me the papers and I'll file the flight plan while you do that." The clerk took the papers and I wrote out our destination and estimated time of arrival. The clerk looked at the flight plan and compared it with the clearance papers we had been given the night before.

"Senor, these are the wrong papers." Handing me back the official papers we had received.

"Why, what's wrong?" I asked suddenly feeling queasy.

"These papers are for another aircraft," he replied. On examination of the papers I saw they had put the wrong N-number for Mortimer. Oh shit, I thought. Harvey had not read them. El Jefe had us in his clutches again.

"There is a mistake here. Let me talk to your boss."

"I'm sorry, senor. He is not here. He will arrive later. You will have to wait." He knew what was going on. He had flashed that official smile of superiority.

"We're in for trouble again, Harvey," I said walking up to the plane.

"Why, what's wrong?"

"Look at this N-number on the document."

"Oh hell! What do we do now?"

"I think we should wait. They may be sitting out there watching, hoping we try to leave."

"We have a full load of fuel and everything is ready. We could take off across the field like we talked about yesterday."

"I'm afraid before we could get Mort warmed up, they would have us in chains." Taking off with a cold turbo is dangerous. Depending on the temperature, the oil in the system must have time to warm up or we could blow off a cylinder from too high internal pressure. Generally, this takes from five to ten minutes. In that length of time they could have us in jail. "It isn't worth the gamble, Harvey. Let's wait."

"Okay," he sighed.

We were definitely working on Mexican time. Nobody was showing up, especially to accommodate us. "We can't call Enrique. Did he say if he was planning on coming back tonight?"

"I didn't hear what his plans were," Harvey answered. "Maybe we should get hold of his friend."

"He didn't give us his business card," I replied wryly. "I wonder why?"

"If we knew his call numbers, we might get through on the radio."

"Probably better not to bother him, anyway. Who knows, he might be in on this, too."

"That is a possibility, isn't it?"

"I don't believe we can trust anyone now. We'll have to wait and play the rest by ear."

"Let's find some shade. It would be cooler flying."

The day inched by. There was nothing to do but sit and wait. It was getting hotter and the bugs had learned there were stationary targets available to harass.

"I'll leave a message with Enrique's wife that we are still here. He might get back and have some ideas," I said getting up from the terra cotta floor.

"What did she say?" he asked as I came back from using the phone.

"He's due back about five o'clock. She will tell him that we are still here."

"Well, that's a plus. My ass is getting tired of this floor."

"At least it's cool tile. All things can't be bad."

The day kept struggling by.

El Jefe came in the main door. He saw us waiting and said, "Ah, Senores, you are still here. We thought you would be gone." As if he did not know, I thought.

"There has been an error in our papers. We have been waiting for you to help straighten it out."

"Oh? Wait here. I shall see what is incorrect." He had one hand in his pocket and kept moving it around.

"That little son-of-a-bitch wants a pay off," I said to Harvey as El Jefe moved away.

"Looks that way. They kept us here all day thinking we have to leave, so we'll be a softer touch."

We got lucky. Enrique came in the door carrying a large gunny sack. "Hi guys. Forgot to give you the corncobs last night. Glad you could wait for them," he laughed. "What's wrong now?"

"They gave us fictitious papers last night. Now it seems like he wants money."

"I'll find out," Enrique said as he walked toward El Jefe's office. Through the door we could see another arm waving discussion.

It was nearing sundown as the two of them came out to the counter. "We have reached an understanding, Mac. You may leave if you wish," Enrique said.

"Anymore conditions," I ask, worried.

"Yes, El Jefe will explain," Enrique winked.

"Unfortunately it is very late and since it is unlawful to proceed after dark, I can only allow you clearance to El Rosario," El Jefe said. Rosario was a little town on the Sea of Cortez. I had been down there fishing once.

"My friend is in charge of the landing field there. He is expecting you. If you go now you will get there before the night falls."

Harvey and I looked at each other and knew intuitively what to do. "Fine with me," Harvey said and started moving toward the flight line.

"Okay, that's a good deal. It will be more pleasant spending the night by the water. Thanks again, Enrique. See you in Arizona."

"Don't forget your corncobs," he said looking at me very seriously.

"Can't forget corncobs," I said nodding that I had recognized his unspoken warning. I tried my best not to run toward the plane.

The engine fired and came to life on the second turn of the prop. Mortimer seemed as anxious to leave as we did. Poised on the end of the runway, Harvey quickly ran through the preflight check, eased the throttle to full power, released the brakes and we were off. The time seemed like an eternity waiting for the oil temperature gauge to reach the green zone.

Mortimer made a sweeping turn to the north as the last light of the sun was disappearing. Climbing under full power reaching for ten thousand feet, Harvey said, "Are you thinking what I'm thinking?"

"I probably am. If we go to Rosario, El Jefe's buddy will collar us again."

"You're right. They are setting us up for sure."

"I really would rather go home. We'll have less trouble clearing Nogales than gambling on being allowed to leave Rosario tomorrow."

"If we made a run for Arizona now, we could be half way there before they figure out we're no-shows."

"I don't see how they would have time to catch us, do you?"

"Especially if they can't see us. Right?" said Harvey. I reached over and switched off all the running lights. "Hey, Mac, that's against the law," he said smiling.

We were able to contact Nogales control from about twenty miles south of the border. We had to make a special request for an agent to meet us in the US. After dark there was no official present at the airfield and someone had to turn on the runway lights.

On the north run, we had kept busy searching the skies for a check plane. There was not a chance for the Mexicans to mess with us now. My thoughts turned to the problems we could have in our own country. "Figured out a story to tell them yet?" I asked.

"They probably won't believe whatever we say. Let's hope the guy on duty knows us. We were fortunate. The agent on duty knew us from other crossings we had made.

"Welcome home, fellows. The bet around here was that we would not see you again, for a long while."

"What do you mean," asked Harvey.

"We've been monitoring your whole escapade since you left," he laughed. "What did you bring back?"

"Not much. Just a sack of corncobs." I handed him the sack.

"Nothing else in there?" he asked holding up a cob.

"No!"

"Listen to me guys. You're not going to be able to pull off this kind of bullshit anymore. Things are changing on the border. Unless you want to get shot down, you had better start following the rules."

"Yes sir," said Harvey. "We learned our lesson at Culiacan."

"The government is putting hundreds of new people down here in an effort to stop the drug runners. These new agents won't know you and are not going to cut you any slack."

"We've noticed things were getting more formal," I responded.

"This border patrol work isn't fun for me anymore. The world has changed for us. I'm going to take my retirement next year."

"We saw some things on this trip that are disturbing to us," I said.

"What you saw is dangerous to your health. You have to be careful. Now, get the hell out of here. I want to go home."

During the flight into Tucson, Harvey said, "We were lucky getting this far."

"Harvey, years ago my dad told me to remember that there wasn't any good or bad luck. It was either good or bad management."

This was our last trip to Mexico with Mortimer. Harvey and I both were given a Sign. We had not managed very well and we would never learn how much money Enrique paid El Jefe for our release.

ARIZONA

After I had gone over the details of our latest escapade, Mary asked, "Well, Mac, what have you learned from all of this? What did you gain?"

"We gained a sack of corncobs and learned that Mexico will forever be an adventure."

"That and a dollar will buy you a cup of coffee," she quipped. "What are you going to do now?"

"We'll hammer up the corncobs for Enrique and hope he can make a deal on the machine when he gets here," I answered.

"When are you going to write the letter to the judge in Chicago?"

"Shit! I forgot about the damn letter."

"Be a good idea to get it done."

"Yes, mother, you're right," I answered wincing.

How does one go about writing to a judge? I pondered, and chewed on an old pencil. How does the opening begin? Dear Judge, no. Your Honorable Sir, no. Maybe Esteemed Highness. How about Your Honorable Esteemed Highness. That probably won't do.

"How are you doing with that, Mac? You should be finished by now."

"I'm doing fine. You can read it in a few minutes." I stared at the page. How do I write this without sounding more stupid than I am?

After a while I wrote:

> Dear Sir,
>
> As per your request, I am submitting what amounts to an outline of the history of our involvement with U.S. Gold. Originally, I was hired

to oversee their mining operations at the Gold Lake Mine in northern California. Subsequently, I took the responsibility of President of Gold Lake Mining Corporation. This corporation has a contract to supply gold to U.S. Gold in return for financing assistance.

The mines that we are developing have had extensive study and have been judged as viable projects. One project alone, the Sierra Pinta mine in Mexico has over fifty thousand dollars invested in time and machinery. Personally I have worked and lived with this property for the past two years.

We have been able to arrange agreements with other mines to aid in supplying the quantity of gold our contract calls for. The mines are all staffed with capable people who, in my opinion, have the expertise to produce the product required.

U.S. Gold has agreed to supply our company with the necessary funding needed to get these projects into full production. To date we have not received much money. The funds we did receive have been paid out for contract development and general mobilization of personnel.

Neither Gold Lake Mining Corporation nor I have any direct involvement with the sales operations of U.S. Gold or their management decisions. I hope this will help you in the matters at hand.

Sincerely,

Mary typed the letter and I dutifully mailed it to Chicago believing there would be no problems with my friend, the judge.

"Everything is in order, Enrique. Mr. Burkett has agreed that you can have the Mexican rights on his patent for ten thousand dollars." I had picked up Enrique and his wife at Tucson International.

"We are ready to put a contract together, Mac. This is a good opportunity for me to get some cash out of Mexico. The peso is rapidly losing its value."

"I need to get the one machine ready to ship to Los Angeles. We have a time limit to start production there. I sure am glad you could come so quickly."

The meeting with Roy Burkett went very well. Enrique came to an understanding with him and had a hand-written agreement with a commitment that a formal contract would be signed later. Roy accepted a personal check and excused himself to go to the bathroom. Enrique and I were discussing a procedure to get the drawings prepared for the unit.

We heard a crash of breaking glass and a thud as if someone had fallen. "Roy, are you all right?" I called out. No answer. Pushing the door open, I saw Roy on the floor. His eyes were rolled back almost out of sight. "Call 911, Enrique. This doesn't look good to me." The ambulance arrived, and Roy was immediately transported to the hospital. They determined that he had suffered a stroke.

"It was fortunate for Roy that you were there when this happened," Mary said.

"Sure was. We saved his life. No one would have missed him for several days. I wonder what else can happen. What do the stars say?" I asked.

"Not good! The rest of the month looks black to me."

"Oh well, I'm getting used to black days. Better get back to LA. If the money has been deposited, we'll arrange to ship the new machine. Roy will be in intensive care for a while. Maybe you can check in on him."

"Glad you're back, Mac," Roger said as I walked into the office. "How did things go?"

"Pretty good, all things considered. We almost lost the airplane, but didn't. Roy almost died, but didn't."

"What?"

"Nothing serious. Everything is back in order. Just business as usual," I laughed. "What's happening at the plant?"

"Jim called this morning. They are at a standstill. They need money to proceed," offered Roger.

"Did we get the deposit?"

"No! There is a big problem brewing."

"We don't need a problem. What happened?"

"Better sit down. I don't think you'll like this." I lit my pipe and thought, it could not be worse than what Harvey and I had just gone through. "I got a rumor that one of the salesmen in Chicago turned over several bars of silver to the SEC yesterday."

"Why did he do that?" I asked puzzled.

"My source says he is making a deal with the government."

"What kind of a deal? Why?"

"I think they have offered him immunity if he helps in their investigation of Harold."

"That sounds serious."

"Very serious. It gets better, though."

"Better? In what way?"

"That silver had been set aside for our use. This was the money that was to be deposited for us."

"How reliable is your source?"

"Reliable. Harold told me."

"Where is Harold?"

"He left town. Wouldn't tell me where he was going."

"Jesus. I guess we do have a problem. What time is it?"

"Ten thirty," Roger replied looking at his watch.

This moment was one I had not wanted to think about. The lurking feeling had been Danny and his company would not get us any more money, but to think any different would have been negative energy expended. Now the reality of the problem was raising its ugly head. "Let's go get a drink!" I suggested. "We have to figure some way to get out of this mess."

"I need two. Let's go. I need to get away from this office. Each time the phone rings it seems to be another disaster. The lounge will be more relaxing."

The lounge at this time of day was quiet and peaceful. "Roger, what is your gut feeling on all of this?"

"I really think we should get out of Dodge and get clear of what is happening."

"But if we do that, we leave all of our people out in the cold," I responded.

"Well, Harold seems to have left us in the cold and what do you think Morales is going to do?"

"I would like to run, but to walk away from all we have built seems kind of chicken shit to me," I said.

"That may be true, but remember I went through the same before. I saw what the government did to Goldstein-Samuelson."

"Well, I can't just quit. There must be an answer for us."

"Okay, Mac. I still think it is time to do something else, but I'll hang around a few more days and see what develops."

I called Jim and Monty at the plant and gave them a rather lame excuse about why we did not have any money for them. Then drove down to Newport to get some ideas from Richard on this latest development.

"Without these funds we're in hot water, Richard. Our first product delivery is due next month."

"Things appear a little black, but since all the contracts are in place and the mines are ready for production, maybe you can find another source of finance," offered Richard.

"We'll have to try. What we have to offer is solid. There must be someone out there who would be interested."

"I have a contact with a retired general who had some interest in silver. I'll see if we can get an appointment."

"We have a short fuse going. Hope he's available," I responded.

Richard answered the phone and with a frown responded, "Yes, he's here, hang on a moment." Handing me the telephone, he said, "It's Chicago. A judge!"

"Mr. MacDonald, I have received your letter of explanation and first I want to thank you for getting it to me in a timely manner."

"Thank you, your honor. How can I help you further?"

"I have reviewed all of the information on this case and feel I can get you clear if you will agree to two stipulations."

"What are those, your honor?"

"One, that you will not deal in securities."

"That's easy. I don't deal in securities anyway."

"Two, that you will no longer be involved in mining."

"You mean that I'm to stop doing what I do?"

"Yes, that is correct. No securities and no involvement in the mining business."

This seemed insane. I had spent years learning about the mining business. I had spent ten or twelve years learning something I had thought was useful and that I loved. I was not ready to give it up. This way of living with nature was what I wanted. To give up a lifestyle because some asshole decided to play the game wrong was unthinkable. A million flashes of red lights from anger seemed to go off in my head. I made a snap decision. "I can't agree to that. This is my whole life."

"I'm sorry. I had hoped you would agree."

"I am sorry, too your honor," I said as he hung up.

"What was that about," asked Richard.

"I'm not sure." Looking back at that moment, I realized the Judge did not care if I went on mining. He was only trying to give me a chance to go away and be quiet. I got another Sign, a big one.

"Roger, we have a meeting set up for tomorrow with a potential investor. Would you like to come along?" We were at our office.

"No, I can't. My wife and I have some problems to take care of."

"Everybody in our group is in trouble. Looks like we have a crisis deluxe," I answered.

"I hope the new prospect likes the program."

"I think we have a strong picture to present. The question will be how fast they can come up with some money."

"Good luck! I'll see you tomorrow night." He waved as we went through the door.

Two days later, I drove out to the refining facility in Saugas.

The Bishop was there. He was concerned about the payment that was past due on his property. As usual, Monty was working in an acid cloud while mumbling over his strange brew. "What are we going to do now, Mac?" Jim asked. "My partner in Susanville put a mortgage on his house to get the mill together. He needs money to carry on."

"That's the reason for this meeting, Jim. I've been working on a couple of ideas.

"How do you feel about U.S. Gold? Will they go ahead with their agreement," the Bishop asked.

"Roger and I feel that should be a good bet. If they do not complete those sales agreements, there will be trouble from their customers. Meanwhile, we must work out a way to keep our program moving ahead. So I'm talking to other investors."

"How does it look with the general," asked Slagle.

"He is very interested in the program. He asked us to call next week and get his decision."

The Bishop interjected, "You fellows are working hard on this project and I know your intentions are right. However, I must tell you that I have another offer on the property. I can hold them off a short time, but remember, money talks."

"This place is our focal point for all the operations, Mac. We can't lose out here," said Monty.

"I know. With the lab operational we can stay in business by buying the gold from Elodia and her father."

"Unfortunately we need cash to do that," Jim offered.

"It's a strange thing," said Slagle. "It takes money to make money."

"That's the shits, ain't it!" offered Roger.

"We have enough set up here to refine the bars from Mexico. Why not make the same arrangement as before. We pay after we sell the refined product," Monty suggested.

"That will work if we can get money to put fuel in the plane and pay Harvey," I said. "It's a vicious circle."

"Alphonse will start pressing for his lease payment for Mortimer, too," Roger reminded us.

"I had thought of calling one of the gold customers, too. Maybe we could raise some funds that way," I said.

Roger blurted,"Now, that's the worst idea you ever had!"

"Why? They're already sold on the idea of gold ownership."

"It's a stupid idea, Mac. We shouldn't do that."

"When can we get Burkett's unit over here, Mac?"

I shrugged and answered with a smile, "When we get some money, Jim."

"We all have a lot of work to do. You are the one to solve the money problem. So do what you think is best, Mac," said Slagle.

"Thanks a lot, Bob. All I can see to do is roll the dice."

"Roll 'em then, Mac. We're all under the gun," said Jim.

"I have to go, guys. Good Luck!" said the Bishop.

"Thanks for your input. We'll contact you as soon as we figure something out."

Roger and I drove out of the plant. I had forgotten to mention the call from Chicago.

Roger was going through the mail. He looked toward my desk and said, "When it rains, it pours."

"What now?"

"The lease on your apartment is up. They want us to renew now or they kick you out." We had never made payments on the apartment. It had been covered on the original lease that U.S. Silver had arranged.

"Man, that was the furthest thing from my mind. How much do they want?"

"First and last month's rent. Three thousand dollars, plus maintenance fees."

"Have you heard anything from Harold?"

"No. He has not contacted me. Have you talked to Alphonse?"

"No, I haven't. Last I knew he was in Arizona working on the garbage deal with Burkett."

The telephone rang and Richard was on the other end. "Mac, good news. The general just called. He said he and his group would make a deal with you. He wants a meeting on Thursday."

"Well, finally something is right with the stars. Thanks for the input. See you Thursday morning."

Thursday went well. We were able to make an agreement with the general and his group, stating they could buy our gold at cost. They would make a nice profit on the resale. Things were looking brighter. I relaxed a little. We would get enough cash from this deal to finish our projects. We called all of the fellows in our group with the news that the dice had come up on seven.

On Monday morning the fit hit our Shan. I was at my desk trying to arrange a way to catch up on our financial obligations when Roger arrived. "Mac, remember those kids you gave an interview to from the Wall Street Journal?"

"Sure, I really liked that pretty little girl."

"Well, look at what your pretty little girl did for you." He tossed the paper on the desk.

On the front page was an article about the purported dealings of U.S. Gold and Harold Goldstein. It described an interesting view of the supposed scheme to bilk the public of all their hard-earned savings. It was mostly bullshit in my opinion. The main problem though was the part that read:

> After an exhausting interview with Eugene MacDonald, President of Gold Lake Mining, it was deemed by this reporter that the authorities should take a long hard look at his involvement with the newest Goldstein caper. In our investigation we learned that he has already been questioned by the U.S. District Court in Chicago.

"How could they do this?" I said looking up from the paper in dismay.

"They are reporters, Mac. I warned you."

"Yeah, you did, Roger."

"This is not good, not good," Roger lamented, running his fingers through his hair.

"Calm down, damn it!"

"There won't be any chance for us now. We're dead."

"Oh, come on. We are in the clear. You know that."

"Mac, I've been through this before. When they grabbed Harold, remember?"

"Let's think about this a minute. Sit down, please. You're making me nervous pacing around like that."

"You had better get nervous. This is disaster, I'm telling you!"

"Roger, I don't see that we have a problem. We can prove everything that we have done. How can they say we're the ones ripping off the public?"

"These people are vultures. They will be all over us like flies on cow shit. I can feel it. I know!"

"We can explain. Somebody in the system will see our side."

"Did the judge in Chicago see our side, Mac?" He did have a point there and his past experience proved correct. We were still arguing about the seriousness of the allegations when a knock came at the door.

"Yes, can we help you?" asked Roger.

Are these the offices of Gold Lake Mining?"

"Yes, how can we help you?" I responded.

"I'm the US Marshal and I have papers here that will allow my staff to collect all the records of the Corporation. Would you make them available, please?" The rest of the morning was like a dream. It did not seem real, but it was. Four or five people started opening drawers and filing cabinets, cataloging the contents and putting the papers into cardboard boxes. After a couple of hours, the Marshall handed Roger a receipt for the records. It was done, and so were we, it seemed. Time for another drink.

"Let's take the boat out and sail around the bay. I think better on the water," Richard suggested.

"I'd like that. Let's go. Any beer aboard?" It was a nice afternoon. There was enough breeze to keep steerage. Richard was at the helm.

"I think you are okay, Mac. With the cash from the general, the project can go ahead. Just don't use the Gold Lake Corporation."

"That's the way I see it, but Roger has gone ballistic."

We were approaching the end of the breakwater when we heard

a bull horn. "Ahoy, there. This is the coastguard. You are ordered to turn back. We have been instructed to keep that boat from exiting the harbor."

"Now what the fuck is this about?" exclaimed Richard.

We tied up in the slip and walked over to the Rusty Pelican. "I'll call my attorney and see if he knows anything about this," Richard said.

"What do you want to drink?"

"The usual. Be back in a minute." I ordered scotch and began to wonder if this could be related to Harold in any way. I could not see how, so I watched the seagulls.

"I don't believe this." Richard was shaking as he lit a cigarette. He had a wild look in his eyes.

"You don't look so good. What did you learn?"

"My attorney had tried to call while we were messing about this afternoon. He was contacted by the attorney for my gray haired lady that bought all the silver."

"What do they want?"

"She wants her silver. And she wants it now!"

"Why now?"

"Some how through the article in the journal, they think I have a connection to Goldstein's deal."

"That's nuts! How could they think that?"

"The SEC took all of your records. What could have been in the files that would connect to Universal Silver?"

I took a long drink of scotch. "Christ, Richard! They probably found the sales brochure on the silver bars. I had one in the file."

"That's probably it." He ordered two more drinks.

"Well, there's no problem. All you have to do is return her silver."

"I can't, Mac. I used it to cast the chessmen for the deal in Salt Lake."

"Can you get the chess set?"

"I'll see. Hope so," he said looking more worried than I ever remembered.

"What was the deal with the Coast Guard?"

"The old lady's attorney views the boat as a potential asset.

They don't want it to leave port."

"Quit worrying, Richard. Maybe we can straighten things out with the general's money," I said.

"I'd forgotten about that. Maybe you could help me out."

"He'll be calling soon." The dominoes had begun to fall.

"Hey, Roger, what's happening today?" I asked as I entered the office.

"More pressure. I'm getting worn out."

Pouring a cup of coffee, I took a deep breath, "I'm a little shell shocked, too. What's the problem?"

"The landlord was here this morning. He wants the rent for the office immediately."

"No chance to negotiate for a little time?"

"I tried," said Roger lighting another nervous cigarette, "but he is afraid of us. He read the paper, too."

"I don't suppose the fact that the government cleaned out the office helped either," I said wryly.

"Have you heard anything from the general?"

"No, damn it! I did not want to call him. It may look like we're too anxious."

"It has been a week since they were supposed to call us," said Roger. "Under the circumstances it may be a good idea to call them."

"I'll get Richard to check on their schedule. Maybe it won't look like we're pressuring them."

Richard called. "Mac, I caught the general on his yacht just as he was casting off."

"That was fortunate. What did he say about finishing the agreement."

"No agreement. They do not want any involvement with us. He read the journal, too!"

"We knew that would happen, didn't we, Richard?"

"Yes, but I didn't want to think about it. I was hoping for a miracle, I guess."

"At this stage we're going to need a big miracle. Well, back to the drawing board. Something has to break the right way for us one of these days. Thanks for the info."

"Mac, I tried to get hold of the people in Salt Lake about the chess set. Their phone has been disconnected."

"That's ominous. What's next?"

"I'll have to drive up there and see what has happened. Call you when I get back."

"Okay, Richard, good luck." Putting the receiver down and slowly pushing the telephone away, I said, "Roger, we're running into blank walls every way we turn. There must be a way out, we have too many good things to work with."

"I don't know how to handle all this. You are the boss, Mac. You figure it out."

Lighting my pipe, while trying not to jump down Roger's throat for coming up with a response like that, the idea came to me to call one of the customers that had already bought a contract from U.S. Gold. "I still would like to try checking out someone on the investors' list, Roger."

"I tell you, it's a crazy idea," replied Roger.

"We might find someone, Roger. The worst thing that could happen is they tell us to get lost."

"I don't like it."

"I'm going to try. Where can we get the customer list?"

Roger called one of the sales representatives he knew and got a few numbers. The first three contacts were secretaries. They would not put me through to their bosses. On the fourth, a brusque voice told me to get lost. He never wanted to hear about U.S. Gold or anyone connected with it. If I wanted to talk to his law firm, he said he would gladly give me their number. "Okay, Roger, you were right, but we did learn something from this!"

"What did you learn from someone telling you to go to hell?"

"That we have no more chance than a snowball in hell of ever finding outside financing for this project," I offered.

"What can we do?"

"Nothing around here. Maybe we can beat the bushes in Arizona. We have to shut down this operation. Now! Maybe if we get away from these crazy people we can see how to solve this problem. Arizona is a quieter place to think, anyway."

"Well I told you we should get out of Dodge a couple of weeks ago."

"Guess I'm not as smart as I would like, Roger."

Monty was busy with his hot acids as usual. "Hope you brought some good news, Mac," he said peering through the vapors of his concoction.

"No, I haven't. We have acquired more problems. When you get a break in your procedure, I need to talk to you."

"I'll be finished in a minute. While you're waiting, take a look at the material in the microscope."

I adjusted the binocular scope. A beautiful mountain of crystalline gold came into view. It took me a few seconds to remember I was looking at a magnification of one thousand times larger than the small sample. "That is fantastic, Monty. Which project did this come from?"

"Susanville. Jim is up there now. That is very high grade ore."

"Boy, I hope there is a lot of it. Monty, you had better sit down. I've got some disturbing news."

"The Bishop told me there was something going on. I talked with him on the phone this morning."

"We aren't going to get the money we have been waiting for. We have to regroup."

"The Bishop won't let me stay here if we can't finish the agreement," Monty responded with a pained look on his face.

"I know. Do you have any room to store this machinery at your house?"

"I suppose we can store it, but there is no place to do any work."

"I'm sorry, Monty, but for now that's the only thing for us to do."

"Damn it! We were all ready to produce, too."

"What did you learn, Mac?" Richard asked. We were in his office in Newport Beach.

"It's harder to shut this down than it was to build it. Bob Slagle is a problem. When he heard what the plans were, he threatened me with a legal fight of some sort. The Bishop was not upset. He had another buyer for his property waiting for us to default. Jim is up in the mountains. I left a message with his wife to call as soon as he gets back. I called Mary. She will bring the pickup and take me back to Tucson. Monty is shutting down the lab in Saugas. Roger is arranging to shut off the telephones and is dealing with the property owner.

"A lot of people are getting hurt, aren't they?" responded Richard.

"Yes, they are and I feel responsible for allowing it to happen. What did you learn in Salt Lake City?"

"I'm sick to my stomach about that. The savings and loan has gone out of business and the chess set has disappeared. I can't handle it any more. I am forced to shut down, too."

"There has to be a way to get back on track. Hopefully, I can find an answer in Arizona."

"Let me know when you've got an answer. My brain doesn't seem to be working right now," Richard said with a tired voice.

"How are you going to handle Mrs. Brown. Any more static there?"

"I went to her house and explained what had taken place. She understood and said she could wait a bit to give me a chance to return her money."

"That sounds good. You only need a little time to recuperate."

"I was able to breathe easier until her attorney convinced her to go ahead and press charges. He advised her not to have any more direct contact with me."

"Damned attorneys. They should all be shot."

"Most of them anyway," agreed Richard. "There must be some good ones someplace."

"I haven't met any yet. I tried to get some advice from the guy in Beverly Hills. He told me he would help us out if I gave him twenty thousand dollars for a retainer."

"That's the shits!"

"Well, we can't quit. We'll fight our way out of this."

There were other people around the world with problems,

too, although I felt all the worries had been dumped on us. Indira Ghandi had been charged with election fraud. President Ford pardoned Richard Nixon. And England had achieved twenty two percent inflation for the year.

Even the trip back home to Arizona was an ordeal. We were forced from time to time to leave the main highway in search of gasoline. OPEC had increased the price of oil seventy percent and reduced production by fifty.

"Seems as if our predictions were correct on the economy," I said to Mary. We had just filled the gas tank and paid one dollar fifty cents for a gallon.

"Being correct isn't much consolation, is it?" she mused. "So, what do you plan to do now?"

"Gold and silver prices are definitely going to jump under these circumstances. We still have the mines. All we have to do is find another place to do the processing."

"There may be a problem with the Pinta," Mary said.

"Oh, why?"

"I don't know exactly. Wilbur called. He wants to talk with you. He said he had some information from Hermosillo that didn't sound encouraging."

"Oh boy, isn't there going to be an end to this?"

"Don't worry, Mac. There is always an end."

At least there was not much traffic on the road. The lack of gas was making it safer to travel.

We finally arrived in Tucson and had a good night's sleep. At least at Mary's place there were no sirens and honking horns.

"Remember to call Wilbur," Mary said.

I was lucky to find Wilbur with a minute to talk."Wilbur, Mary says you have some news for me from Mexico."

"I do. I have a friend who is a top aide to the Governor of Sonora. He called and told me there had been a directive sent to the customs in Lukeville to pick you up. It could be there now."

"What is this about?"

"Apparently because of the problem in Los Angeles, Montgomery thinks he has an opening to take over the mine."

"What has he done?"

"Your good friend and partner swore out an affidavit claiming that you had been smuggling machinery and dynamite into Mexico without paying the duties. On top of that he swears you have been stealing gold from him and smuggling it back to the states."

"That dirty son-of-a-bitch!"

"The recommendation from my friend is you stay away from the border. Don't try to go down to the mine."

"You mean that I can't even go and argue with Montgomery?."

"If you get caught, they will put you in jail, Mac."

"He is stealing everything we have there, Wilbur!"

"Since you're in trouble here, you won't stand a chance down there."

"I'll kill the dirty bastard!"

"You won't have the chance unless you can catch him on this side of the border. Looks to me like you are fucked. What are your options? You can be put in a Mexican jail for smuggling or a penitentiary here for murder."

"I would pick the penitentiary. At least that miserable shit would be dead!" I was pissed. After the two years that I had spent getting that place organized and making it a workable project, to have one cross-eyed bad American appropriate everything for himself, was a little too much to handle. He had not even helped me do much of the work.

"Better sit down and have a beer. Don't kill any one just yet," suggested Wilbur.

"Goddamn it, this is impossible. Everywhere I turn there is a block wall."

"Drink your beer, Mac," advised Wilbur.

"Sure I will, maybe two. Thanks for the input," I said, hanging up the phone.

"Damn it, Mary! What in the hell is next? We lost the refinery, the Pinta and we don't have any way to get more financing. I do not see how we can hold on to the dredge in Alaska. Clay and Jim Scott are sitting out there with good mines to work and they can't move until I send more money."

"Another problem you are forgetting, Mac," Mary said as she dried another dish and reached toward the cupboard.

"What else?"

"Alphonse Morales wants his payment on Mortimer."

"So, he's going to pressure us too. Is everybody against me?"

"No one wants to be associated with the poor."

"We can still make use of the plane. We have to be able to get to Mexico and Susanville. They are too far to drive. How can we keep Mort?"

"I'll talk to Morales, Mac. Maybe he'll deal with me."

"I'm not so sure. I think he likes boys."

"Better for me if he does. I've got an idea on how to save Mortimer."

"Okay, you work on Morales. I've got to find a way to raise some cash," I said. An idea popped into my head about the possibility of using the rock grinding machine. Maybe I could locate a place to put it to work. I remembered the fellow from Colorado who had been at Roy's grinder demonstration that Elodia and I attended on our way to LA. I found his card in my wallet.

Luck was with me. A secretary found Bernard in a couple of minutes. "Mr. Mienerz, I met you at the county yard in Tucson. You asked me to call you if I was going to be in your area."

"I remember you. How are you doing?"

"Pretty good, thanks. I have the rights to the pulverizer you saw and wondered if you had thought of a way to use it?"

"Actually, I have. We have an application for it in one of our plants. We are grinding silica for Coors and need a better way to do it. Could you come in and we can discuss it?"

"Sure, I think we might be able to get up there first of next week."

"I plan to be in all week. Give me a call a day or so in advance."

"Okay. I look forward to meeting with you. Thanks, Mr. Mienerz."

"Just call me Bernie. See you in a few days."

I called Richard in Newport Beach. "Richard, I found a guy in Denver that is interested in the grinder. He wants me to come up and talk about using the grinding machine. Could you get away from LA to help me with this?"

"I can in a few days. I'm in the process of liquidating, selling what I can and giving away the rest."

"Save your drawing set. We may have to design a flow sheet for these people. If we can get this job, we'll at least be able to buy some groceries."

"What is it they want to process?"

"Silica. They have a contract to supply Coors. They must use it to make their bottles."

"That sounds interesting. Are you planning on flying up there?"

"Yeah, if I can find Harvey."

"Hi, Harvey, what are you up to?" Harvey was underneath his old Ford.

"Trying to replace a U-joint. This old car ain't going to last much longer. Hope you found some money. I need to get some parts."

"We have a shot of developing a job in Denver. Can you get away for two or three days?"

"I won't fly Mortimer until you get some maintenance done."

"What's wrong?"

"The throttle sticks and it's hard to adjust the fuel mixture and according to the log, its time for new spark plugs and wires."

"What would you estimate that would cost?"

"I'd guess between three and four hundred dollars."

"Where are we going to find that much money? Can we get a maintenance manual, Harvey?"

"Sure I know of one I can borrow."

"Let's go buy the parts. I'll fix old Mort."

"That kind of work is supposed to be done by a certified mechanic, Mac."

"Will the mechanic fix it for nothing, Harvey?"

"Of course not!"

"We have enough money for the parts. Get the book and I will repair the airplane. Mary arranged to take over the contract from Morales. Since it's her plane, she won't mind if we save some money on the maintenance."

"That reminds me, Mac. I saw Dick the other day. He is still pissed about you getting his flying machine."

"Why? He got paid for it."

"He's jealous. He doesn't like the fact that you have it."

"What an asshole. Dick is one more nice guy on our list we do not want to do business with."

"Yeah, he is a nut," agreed Harvey. "Have you heard anymore about the government investigation on Goldstein?"

"No, not a word. I think the whole thing has dried up and blown away. We'll have to work out a way to keep going on our own."

The new rules spelled out by the modern airports in the US did not allow Mortimer any closer than twenty miles to the flight patterns. Mort's radios were just not good enough. To upgrade his electronics to the specifications of the FAA would cost more than we paid for the complete aircraft. Therefore, we set down at the Broomfield Airpark on the outskirts of Denver.

Bernie Mienerz had sent a car to pick us up. "How was the flight, guys?" the chauffeur asked.

"Not too bad except we collected a lot of ice near Taos and this eighty knot wind you have here made us pucker the seat cushions," answered Harvey.

"This business of flying in the mountains of Colorado is different from northern Mexico, isn't it Harvey?" I observed.

"It's better here. At least there was no army to meet us," he laughed.

"Where do we go from here?" I asked.

"The office is right at the mousetrap, about forty minutes."

"Let's go before it gets too late."

Bernie owned a yeast plant, here in Denver. They used animal blood from the slaughter houses to render the yeast. It was obvious Bernie was a big game hunter and his office doubled as his trophy room. A big Kodiak bear was standing upright, snarling down at us. The bear was at least eight feet tall. On the walls were heads of bighorn sheep.

"This could be a good opportunity for you fellows," explained Mr. Mienerz. "You bring the machine up here to Colorado and set it up at our plant. We will run a test with the product. If the grinder does the job, we will need several of them."

"Where is the plant located, Mr. Mienerz?"

"In Salida, about one hundred fifty miles southwest of Denver or fifty miles west of Pueblo. Here is a check to start the project. Get your machine up here and go to work." Obviously this guy had not read the Wall Street Journal.

On the return flight to Tucson Harvey said, "Mac, I'm having trouble with my wife. She is pressuring me to get a paying job. I am not going to be able to fly with you anymore."

I had been flying Mortimer from the right hand seat while Harvey relaxed with a cup of coffee. "Women are a problem, aren't they? How are we going to get along without you?"

"You have a lot of hours in this airplane, Mac. Why not apply for your license. I know you can handle Mort if you stick to normal airports."

"I'm feeling more comfortable each time we make a trip, but it's nice to know there is a backup sitting next to me," I answered.

"Apply for the license, Mac. All that is required is a physical. When you do that, I'll show you how to do some emergency procedures and sign you off as proficient."

"Seems like a good plan. Can you help us for a few more days? We'll have to get both the machine and the plane to Colorado."

"Sure, I can do that, but I'm still forced to apply for a full time flying job."

We refueled Mortimer in Tucson. The cost of filling the tanks had risen about forty dollars. Everything we did, was costing more and more. The race was on for us to get our gold projects into production before the price of machinery and fuel put development out of our reach. There was no time left for any more mistakes.

"Mary, this doesn't look bad for us. If we can get the test done in Salida, there will be some money to keep the mines alive."

"Well, it is encouraging that they put some money up front. Perhaps you have found some real people to deal with."

"Mienerz looks solid. He owns a dehydrating plant in Denver that manufactures yeast. He also has the silica project for beer bottles at Salida."

"What is your plan, Mac?"

"We'll load the machine on a trailer and take it up there with Merlin. We have to get someone to drive up there because I am going in first to find a place to stay and get an idea how to set up for the test."

"I'll drive up," offered Mary.

"Who's going to drive the school bus if you're gone?"

"I quit the school. Also, I have sold the place here. That's where the money came from to put the down payment on Mortimer."

"You quit and sold your house? What are you planning to do?"

"I'm tired of this job, Mac. I have decided to go with you. I'm interested in what is going to happen next with this saga of yours."

"I don't believe it. I did not know you were crazy, too! It will be nice to have someone around though. The snow is deep and cold up there. Maybe you can keep me warm."

"I'd like to keep you warm, Mac."

"What will your daughter think?"

"She eloped didn't she? Why can't I?"

"Your daughter thought she had a reason!"

"I have a reason but it is not pregnancy!"

"What is it?"

"I like you, Mac. I want to be with you."

"Now I know you're nuts."

"I'm not crazy. Let's go to work. I am anxious to see what happens in the next chapter of your unorthodox existence."

"Okay, you are the designated truck driver but you might have to be the head cook, too."

"Deal," she said. "Give me a kiss!"

As I hugged her I thought, now I know why I had not heard from Elodia. Mary had hidden the letters. There can be deviousness along with honesty.

COLORADO

We followed the Arkansas River up from Canyon City and found Salida nestled in a pretty valley. The landing field at Salida was on a mesa about seven thousand five hundred feet above sea level. The southern approach was over a bluff that dropped straight down about two hundred feet. "Okay, Mac, let's see you land old Mort. The only thing I can see that might give you a problem is a possible updraft at the edge of that cliff. Keep your approach speed a little higher than usual."

I banked Mortimer to the left and lined up on the distant asphalt runway. I throttled back and adjusted the fuel mixture. Harvey was talking on the radio announcing our arrival to ground control. This is going to be a snap, a straight field and no rocks. As we crossed over the precipice, Mort bounced from the turbulence and gained a little altitude, but this did not present a problem for me.

Glancing at the instruments and making sure of the flap setting, I felt this approach was simple in relation to the ridiculous landings we had been making in Mexico. Everything was going smoothly according to plan. Richard was in the backseat reading. We were about forty feet off the runway and I was just beginning to ease the stick back to flare out when Harvey tapped me on the shoulder. I glanced at him and saw him pointing to a spot ahead on the left. Looking in that direction my heart tried to stop. A herd of deer were trotting out of the brush and crossing exactly where I had planned to touch down.

Under these conditions there is no time to think, only time to react. I hit the throttle and reached for the flap controls. Harvey's hand was already there. Without a turbo charged engine at this high altitude we would have never made it. Old Mort growled as if in disgust and roared just above the deer. Five feet lower and the local townsfolk would have had a lot of venison for supper.

"Damn it, Mac! You made me lose my place in the book," Richard said.

"The deer didn't seem to care. What's wrong with you?" I said trying to act casual.

"Good job, Mac. You did exactly the right thing. Go around and try it again," said Harvey.

"I'll go around, but you put us down, Harvey." My knees were shaking so bad I could not have pushed the pedals hard enough to stop us no matter how long the field was.

We checked into the Palace Hotel in downtown metropolitan Salida. The population count could be stretched to about thirty five hundred if you counted the deer. "How do you like commanding an airship?" asked Harvey as I started my second Coors.

"Yesterday I couldn't spell pilot. Now I are one." My legs were beginning to gain some strength.

"Seriously, what do your think about flying?"

"I think all private flying is a lifetime of boredom and an instant of terror. Might as well work for the CIA. It must be about the same."

"You'll be okay. Just remember that you never stop flying an airplane 'til you shut down the engine at the terminal."

"Thanks, Harvey, wish you could stick around." He was right. I had relaxed too soon on that approach and had let up on my concentration of the problem at hand. I would not do it again.

Salida, Colorado, was a decaying town. It seemed the sheer weight of its history was too much of a burden to keep progressing. The Denver and Rio Grande Railroad's arrival had allowed Salida to thrive and prosper for the last ninety years. Now the railroad was only using their yards for a side track storage of a few ore cars.

"What does Salida mean, Mac? It's a Spanish word isn't it?" Richard asked over his coffee.

"It means a way out or exit. The Arkansas River Canyon is the only outlet to the east from the Rocky Mountains for a long ways, both north and south. Originally, the Spaniards controlled the whole area because of the gold and silver mines. Salida should be pronounce Sah-Leeda." The surrounding towns had Spanish names. The local language had fractured them all. They pronounced Buena Vista as Bewney. Shocking.

Across the river we could see a circular red brick building. "I bet that was the roundhouse where the locomotives were maintained," Richard offered.

"That's a big building for sure. This must have been an interesting place in its day," added Harvey.

"Let's walk over and have a look. Mary won't be here for a few hours," I suggested.

As we were picking our way across the tracks, we noticed another building near the old roundhouse. The windows were broken and through the empty frames, a white smoke was rising in the cold mountain air.

"What the hell, is that place on fire?" exclaimed Richard.

"Sure looks like it."

We peered through the front door and to our dismay, realized the smoke was dust. We had found the silica processing plant. A form was moving toward us through the cloud. It looked like a big polar bear. "Can I help you fellows?" the bear said. This guy was encased with white silica powder. "How in hell can you breathe in there?" I asked.

"You get used to it. What can I do for you?"

"We're looking for Phil. We have a machine to install in this operation."

"Oh, yeah. You must be the engineers from Arizona. I'm Phil. Welcome to the dust bin from hell."

"This damn dust is bad news. Can't be very good for your lungs, Phil."

"That is the reason you are here. Bernie thinks your method of grinding will cut down on the dust problem."

"Show us where to unload. The unit should be here soon."

Walking back to the Palace, we passed a building where the Chinese miners used to smoke their opium. The town folk called it the 'cribs'. Peering into the dark interior, in my mind I could almost see a bank of bunk-beds setting around the room. After finishing the railroad, the Chinese had stayed and gone to work for the mines around Leadville and here they established restaurants and hand-wash laundries.

"Looks as if we'll be here longer than we figured, Richard. That plant is one big problem."

"Maybe the hotel will give us a break on a monthly basis. Let's ask."

We negotiated a rate of sixty-five dollars per month for a two-bedroom suite looking over the park. "Guess they aren't overwhelmed with advance reservations," Richard said pocketing the receipt. He set up a high-class engineering office in our suite and used the kitchen table for the drawing board.

We had put Harvey on a bus back to Tucson. "The way these conveyors are placed is a problem, Mac," Richard said pointing at a drawing.

"We'll have to build a frame to get the grinder machine high enough to reach that feed conveyor, won't we?"

"The problem will be lifting it that high," said Richard.

"The cranes are still there. Maybe we can get them to work. They used to pick up a locomotive boiler. The weight of our machine shouldn't hurt those cranes," I said jokingly.

The electric overhead cranes had not operated for years. We climbed high up in the trusses and found the control box. The only thing wrong was it was full of silica sand. In ten minutes time we had a fifty-ton crane at our disposal.

"That solved that problem, Richard. Let's start building the base."

We worked about three weeks getting set up so we could start our production test. We finished tying in our power cable about two o'clock on a Friday afternoon.

"Phil, we're ready to try this thing out. Can we run a little trial test?" Richard asked shaking the dirt out of his hair.

"No. I have to leave early today. Let's do it Monday morning."

Monday seemed to sound all right. We were tired anyway. We agreed.

At the penthouse suite, the lingering odor of ham, eggs and fried potatoes with onions was in the air. Mary was doing her best to keep us healthy by preparing good food. "You guys should take a day off and see the sights around here," she said.

"Guess we could. Where would you like to go?"

"Leadville. We could check out the molybdenum mine that Climax is operating."

"Mary is a different kind of girl, Richard. Most women want to go to the big city to shop."

"I think someone is at the door. I'll check," she said as she took a swipe at me with the dishtowel. She returned with a fellow who looked like a miniature John Wayne, the way he sauntered in his boots and the tilt of his hat.

"I'm Norman Cochran," he said taking off his Stetson. "Heard you fellows had a miracle grinding machine. If you have time, I'd like to talk to you about it."

"Good morning, Norman. I'm Mac. This is Mary and the guy with the big coffee cup is Richard."

"Glad to meet you," he said while shaking hands. This was the first person in the town that had seemed friendly. The locals had treated us like aliens from another world.

"What do you have in mind, Norman?" I asked.

"Our company is building an ore processing plant up in the mountains. We have a jaw crusher there, but no secondary crushing machinery. I heard your machine was designed for this kind of work."

"Yes, it is, but we have a contract to finish here at the silica plant," Richard said.

"Where is the mill?" I asked.

"South about an hour. The place is called Bonanza. Would you like to come and look? We have time to drive up there today."

"Why not. Sounds interesting. Let's go."

Norman drove south to Poncha Springs. "In the old hotel over there they have a guest register signed by Ulysses Grant. If you turn right, the road goes over Monarch Pass to Gunnison." We turned left and began to climb higher into the mountains. "You can see the old railroad bed. They ran the narrow gauge trains up here to bring out the ore from the mines. The engine in the park in Salida was one of the original locomotives." The roadbed for the track ran along beside the highway and appeared to be a steeper grade than the road we were driving.

"What did they mine at Bonanza?" Richard asked.

"Lead and silver. The ore was rich enough to be shipped directly to Leadville without processing. The railroad went out of business when they built the roads into the back country."

We drove into the San Luis Valley. The Sangre de Cristo Mountains were on our left. "There are several Spanish gold mines up in those hills," Norman explained.

"What's the elevation here, Norman?" I asked.

"About eighty five hundred feet. The mill at Bonanza is ninety two hundred."

"Damn, that makes it hard to breathe," Richard said.

"Naw, you get used to it after a few weeks. Keep your eye out for elk and antelope, there are usually some feeding in this area."

Norman turned right from the main highway and followed a dirt road that led us higher. Soon we were driving through a forest of aspen. "The trees are called quakies around these parts," he explained, "because the leaves always seem to be moving."

"The trees are pretty with that white bark. What do they use them for?"

"Makes good firewood. They don't make much ash."

"I imagine it gets fairly cold around here," said Richard.

"Really cold. It's about twenty degrees out there right now. The town of Bonanza is up the road about a mile from here," Norman said, as we turned off the dirt track and crossed a little bridge. Ahead was a green building rising in steps up the side of the mountain. "We're on our property now. We bought this mill building and one hundred sixty acres for our operation."

We could see several vehicles and pieces of earth moving equipment parked around the building. The first thing I noticed was the quiet. "There are a lot of elk here and higher up you can find big horn sheep," Norman said. "Come in the trailer. We'll get some coffee."

"Richard, this is really nice. Wouldn't it be great to be able to work up here?"

"Did you notice the power line? They must have plenty of electricity," he said.

"That would be novel. All this and electric lights, too. This is a little different than Mexico."

There were four or five workers sitting by the stove inside the mill building. Apparently this place had not been used for several years and the men were cleaning up the rat shit left by the most recent tenants. "From the size of the droppings there must be some big rats around here, Norman."

"Yes, they are big. Trade rats. They scare the hell out of you when they run across the floor."

The building was constructed about 1950. It was designed as a flotation plant to process ores from the surrounding mines. The mill building was only about sixty feet wide. It stepped up an incline on several levels about two hundred feet. Each level was connected by narrow wooden stairways. At the highest point behind the secondary ore bins was the crushing room. The ore from the top fed by gravity down through the various processing stations. All of the old machinery had been removed except for a rod mill.

"Someone stole the pinion gear from that mill. We can't use it." Norman explained.

"Our machine will replace it anyway," I puffed. We were all out of breath just from climbing two flights of stairs. A smaller room to the left had been the testing laboratory. It was still full of rat shit and smelled terrible.

"You have a lot of work to do here, Norman. Do you have the financing to put all the machinery in place?" I asked.

"We have some investors from Colorado Springs and one doctor who lives in Maryland."

"Who's the chief engineer?" Richard asked.

"I am," Norman answered straightening up. I expected to see him adjust his imaginary gun belt. But he did not.

"What kind of machinery do you plan to use?"

"I'm not sure yet. I don't know what ore we have to run. Excuse me a minute. I've got to tell the crew what to do next."

When he was gone, I said, "Richard, this guy is nuts. He says he has a budget to put this mill together, but doesn't even know the type of material he has to process through here."

"Yeah, too bad. I kind of like the guy."

"Norman, who owns the dozer and loader out in the yard?" He had appeared from the lower level.

"The company," he answered. "They have made down payments on all of this. It's getting late. We had better start for town."

By the time we returned, the shadows were beginning to lengthen over our park. Mary asked, "What did you think of Norman's project?"

"It's a pretty place. You would like it," I said.

"There's something though that doesn't feel right," Richard said.

"Yeah, there's something about John Wayne, can't put my finger on it yet.

"John who?" they both asked.

"Never mind, we have our own problem to solve across the tracks."

"Hand me another Coors, please," Mary said.

Monday morning was supposed to begin as a red-letter day. We were on schedule. After a few production tests, we would have an order to build several more machines for Bernie. "Let's drive to the plant, Richard. It's too cold to walk." Along the edges of the river, we could see ice that had built up in the quieter pools of water.

"Looks like we have company, Mac. There are more cars here than usual," Richard observed.

"Maybe they came to see the demonstration." As we walked into the plant, we saw Phil surrounded by four men. None of the machinery was running. It was too quiet.

"What's going on?" I asked one of Phil's helpers.

"These guys are Occupational Safety and Health Administrators. I think they are telling Phil to shut down the operation. We make too much dust, I guess." OSHA was the federal government protection agency for workers.

"Oh, shit! This doesn't sound good, Richard."

"Maybe they are just giving a warning."

"I don't think so," the helper said. "They gave us a warning already, about a month ago." There seemed to be more of a chill in the air. Another black cloud was beginning to form.

"Phil, are you saying that we won't be allowed to even run the test?"

"That's right, Mac. We can't operate until we get a dust control system installed."

"But that's what we set up this machine to do! This test we are ready to do right now, will be part of the solution to your dust problem."

"I know it and you know it, but we have been ordered not to operate."

"That's a Catch 22 isn't it! We have to run the machinery to prove you are solving the problem, but they won't let us run anything until we prove the problem has been solved," Richard swore.

"That's about it. The government thinks a bit different than we do."

"Well, goddam it, Phil. This puts us out of business. If we can't prove the machine, we can't get Bernie to build more."

"I believe we're all out in the street, Mac. I'll lose my job here, too.

"How long will it take you to build a dust control system, Phil?" Richard asked.

"My guess is about six weeks, but that will depend on Bernie. If he thinks it will cost too much, he'll probably shut down this operation and move some place else."

"They could at least give us a chance to show them what we've done toward cutting down the dust."

"You can call President Ford, Mac. Maybe he will listen."

We did not call Gerald, but we were able to get through to Bernie. "Is there anything you can do with this problem. Maybe we could run at night when nobody is around," I suggested.

"No, Mac. If they caught you, they would lay a hefty fine on us. We can't take that chance."

"What are we supposed to do? We can't sit here for six weeks."

"I'm afraid I'll be forced to close that plant. The demands that OSHA has made are too harsh. We may not be able to justify the cost it will take to reach compliance."

"That would mean we would have to take our machine out, then. When will you reach a decision, Bernie?"

"By the end of the week. Phil will get me some cost estimates and I'll decide from those."

"We can't wait too long, Bernie. We're kind of between a rock and a hard place on this."

"I'm sorry, Mac, keep in touch. We'll work something out for you, somehow."

"Thanks, Bernie, appreciate that."

The wind was out of our sails. We were dead in the water.

"Well, guys, now what? The rent is due on the apartment and another payment on Mortimer isn't too far away," Mary said.

"I don't know. The whole plan was based on Bernie's deal. I was so sure I could get this program finished, I had never thought about a Plan B. Whadda you think, Richard?"

"We certainly can't wait for six weeks of refit at the silica plant and we can't go back to Arizona. Seems like we are getting a lot of interest in the machine around here. Maybe we should pursue that."

"We'll have to find a place to get it set up. If we can demonstrate its grinding capability, it will be easier to sell."

"I think we can improve on the design, too. The more I study the principal, the more I understand the physics of the internal movement of the rock particles."

"Norman wanted to try the machine in his mill," Mary said. "Maybe you can arrange something there."

"That's a thought, but that place is a long way from civilization."

"We're already a long way from civilization. What's a few more miles."

"That's a point. Let's drive out there and pay more attention to what we could do with that place."

"It's only fifty miles, Mary. Want to come?"

We were driving south on the main road out of Salida. This north end of San Luis Valley is where the farmers grew hops for the Coors brewery. Our road to Bonanza was a turn off toward the west.

"This is the road Norman turned on, isn't it?" Richard asked.

"I think it is."

"Is that a town ahead of us?" asked Mary.

"I don't know. We did not drive up that far."

"Let's go see what it looks like. It's close," Mary said.

Passing our turn-off, we saw a sign announcing "Villa Grove, one mile". There were no sky-scrapers breaking the skyline. "This place can't be very big. Why have a city way out here?" Richard wondered aloud. We slowed, saw some buildings on both sides of the highway and then there was nothing but pavement stretching ahead.

"That was it? That's all there was?" said Richard.

"I guess that was Villa Grove," Mary offered.

"Here's a place to turn around. No point in going on."

"Watch for traffic cops on the way back, Mac."

"You guys keep an eye out. I'm fighting this heavy traffic problem." We had seen only two cars since Poncha Springs.

"What's the population of this metropolis, do you suppose?"

"I counted six buildings," said Mary. "I saw a post office and an antique shop."

"I saw a liquor store and a bar. Wonder if they have a pool table."

"There were two restaurants," said Richard, "and a phone booth."

"Hell of a place. I didn't even see any people." The town was forty miles from the grocery store at Poncha Springs.

Norman was leaning against the mill building smoking a cigarette. His Stetson pushed back on his head. "Welcome folks. Didn't expect to see you around these parts."

"We were given a few days off. Thought we'd visit God's country," I said.

"How many people live in Bonanza?" Richard asked.

"Eight, during the summer about twenty. This is a resort area you know," laughed Norman.

"What do they do for a living," Mary asked.

"Well, let's see. There's an old guy and his wife. He's the meter reader. A young lady, Gayle, she's an artist. Bob, he is a retired gun dealer. A couple of prospectors, old Bill is one and an old lady who has been here since the thirties. She has bright red hair, buys and sells mines. Her name is Mary Taylor."

"Sounds like an interesting place," Mary commented.

"It is. They take turns each year playing mayor. Gayle is the mayor this time around."

"Is there a saloon?" I asked.

"No. The only thing in town is the hand pump on Main Street. Everyone gets their drinking water there."

"How many streets are there?" Richard asked.

"Two, and a garbage dump."

"Are there any phones?"

"Two telephones. The meter reader has one and the mayor has one. They move that one each year to the home of the new mayor."

"Can you use their phones, Norman?"

"It's not a good idea if you're making business calls. It's an eight party line and every one down the creek will listen in."

"I bet Myrtle is the operator," joked Richard.

"As a matter of fact, her name is Myrtle. If I have to make an important call, I drive down to the phone booth at Villa Grove."

"That's what you could refer to as a long distance call," I laughed. "It's several miles to the phone booth."

"Great spot to live," offered Norman, "not much stress around here. How is it going at the silica plant?"

"Not good. OSHA forced them to stop operating. They may not start up again."

"That's too bad. Hope they don't find this place. What are your plans now?"

"This problem came up so fast we haven't had a chance to decide what to do about it."

"Why not move out here? You can set up the machine on the

second level. The boys are about finished cleaning up. Come on in the house. Let's work out something."

We were back in our suite at the Palace. "Well, what do you think guys? This deal that Norman offered could give us a chance to regroup."

"At least we would have a place to stay. We can't afford the sixty-five a month here at the Palace," Mary said.

"There is enough power in the mill to run the machine. Norman told me an electrician would be up there next week to finish wiring the electrical panels," Richard said.

"I think we got lucky with this. We can use the place to work on the ore from Mexico and handle the concentrate from Jim's place in Susanville."

"It's a long way from where Clay and Jim are working, Mac,"

"I know, but we have a chance now to finish the other program. I should talk to Jim and see what he thinks."

"Norman said he would pay us for any engineering help we can give them and we do get free rent," Mary said. "I like it up there. My vote is to go ahead and move in."

"We really don't have another option at this point. I'm for it, too."

"I'd like to meet the artist. I'm for it," Richard voted.

"Agreed! Open another Coors. We'll toast to our move to Bonanza.

"Richard, I've been thinking about the business world. You know we're getting to be experts at managing things."

"How's that, Mac?"

"Well, we're doing things a lot faster than we used to. It took me about two years to manage losing the Sierra Pinta, right?"

"Yes."

"It took you a little more than a year to get out of the silver business." Richard glanced at me, puzzled. "The maximum time spent on Goldstein's deal was only about three months and now we have managed to get blown out of Bernie's project in a maximum of six weeks. Seems to me we qualify as experienced management people."

"You know what I feel about all that?"

"What?"

"I feel paranoid. I hope this is a real deal with Norman."

"Aw, come on, Richard. That's a bit negative isn't it? What are the percentages of our finding another deal that can fail as quickly?"

"With the history you just outlined, our chances would be about fifty percent less time than we spent on Bernie's project."

I awoke to a quiet stillness that was almost overwhelming. "This is one of the reasons I like these projects," I said aloud. To hell with the Beverly Hills of the world, let the promoters have it.

Mary entered the door of our one room cabin. "What have you been doing?" I asked.

"Cleaning the outhouse. Damn men don't have any class," she said standing a broom in the corner.

"I thought it was a nice place. At least it doesn't have a rattlesnake in the hole," I said.

"No, but those big rats like it. What a mess!"

"I like it because of the two holes. Instead of reading the Sears catalog you can talk to the girl sitting next to you," I offered.

"I don't need that kind of intimacy," Mary retorted pouring a cup of boiled coffee. "What are you going to do today?"

"Finish up the lab. I'm about ready to move in and start some work."

Norman came in the door. "Mornin' folks. Did you sleep warm enough?"

"Sure did. The saw dust in these walls makes a good insulator."

"I've got some news from my investors. They want to know if you would go to Idaho and make a report on a silver mine they have. They'll pay you for it."

"What type of report do they want?" I asked.

"A feasibility study. They're not sure about spending more money on the project."

"How soon do they need the job done?"

"Within a month."

"We could fly up there, but I have get Jim Scott first. He's in California, close to Lancaster."

"There's time for that, Mac. I'll tell my people you'll take the job and have them cut you a check."

"Sounds good Norman. Have some coffee."

"You know what Mary? I think old Mort can make it to Lancaster nonstop." I was studying the Cessna manual. "If there are no head winds, he should have enough fuel to do it."

"What margin would you have?"

"About thirty minutes if my calculations are correct."

"And if they aren't?"

"I'll have to stop in Barstow or somewhere. It will save about an hour if I can go clear through on one hop."

"Is it worth it to cut things that close, Mac?"

"Probably not, but it would be interesting to see if the factory specs are correct."

"What's the flying time?"

"Five and half to six hours if the conditions are right."

"What's the point. Why stretch things?"

"I don't know. Would be interesting I guess."

"Well, you're the captain."

Actually, there was no reason for me to try this bit of foolishness, but I was looking for a way to make a long and boring trip more interesting.

Mort had to climb to twelve thousand feet to clear the Rockies at Monarch Pass, and he was happy to be back to work. The air was smooth at this early hour. He seemed to understand that we were going to try this nonstop. We passed Durango and headed west just north of Shiprock. In the distance, we could make out the Grand Canyon.

Time flies when you are having fun. My fun was figuring the calculations on the distance between airports, and checking on how the fuel supply was holding up. At this altitude, the turbo was using about fifteen gallons an hour. Everything was perfect for Mort, but we were too high for a pilot without oxygen. I had been concentrating so hard on the mathematics of Mort's performance that I had not noticed the change in mine.

Mort, you have done it. There is the control tower. I picked up the mike. "Lancaster, this is Cessna N one six four seven on downwind. Request landing instructions."

The radio crackled a moment, "Read you one six four seven. What is your location?"

"Right next to you," I could see him leaning over his mike in the tower.

"Jeez! What are you doing so close! Okay, cleared to land on niner right," acknowledged Control, "one six four seven out."

Mort made his turn to the right, but because we were so close to the run way we had to negotiate an 'S' turn to get lined up on niner right. All seemed perfectly normal as we touched down, but they were not normal for Mort.

Glancing at the airspeed indicator I saw 120. "Doesn't seem that fast," I said aloud. "Here's the off ramp, Mort, we turn here." I made the turn way too fast and we skidded to a stop off the pavement. The cloud of dust was beginning to settle when the radio said, "One six four seven, are you okay?"

I clicked the mike, "I guess so."

"There was another mile of runway you know. Next time try to use it, please."

"Where do I park this thing? Over."

"Front of maintenance shop. Better check out your tires."

"Sorry Mort. I guess that wasn't a textbook landing. Didn't mean to embarrass you." Mort snorted and took us to maintenance. He had twenty minutes of fuel left.

With Jim and his textbooks aboard we were approaching the Colorado River flying at a more reasonable altitude of eight thousand five hundred feet. I had just finished outlining the possibilities we had at the Bonanza Mill site. "So there you have it, Jim. How does it sound to you?"

"It's all good, Mac. There is one thing, I've been holding back. I'm not sure how you will take it."

"Try me. Can't be too shocking."

"My partner in Susanville had a heart attack. He died two weeks ago."

"Shit, Jim, that's too bad."

"Yeah, he was a good guy, but there's something else. He had mortgaged all the property to keep us going." His voice trailed off.

"We can still get the ore though. Help pay back the mortgage?"

"No, we can't. Everything is in probate and his wife won't let me touch a thing."

"So that means all the machinery and tools are locked up for a year?"

"Fraid so. All of our work there is down the toilet. His wife didn't agree with spending the money on the mine. She'll auction the works when it is clear of the court."

"That's a pisser. Our options are getting more narrow every day."

"We'll get going, Mac. I've learned Monty's recovery methods. We can set up his procedure in Colorado."

"Lucky we stumbled on to the opportunity. Know anything about Riggs, Idaho? We have a job to do up there soon as we get your things organized."

"Yes, your Uncle Steve and I did some studies around the area. No big deposits that far south in the state. I've been meaning to ask, have you heard anything from the Goldstein fiasco?"

"No, not a word."

"That seems a little bit strange," said Jim.

We landed in Durango for fuel. The knowledge that you have an hour or more of flying time was a much better feeling than what I had just done.

Mary was on the roof of our sawdust house. "Just what is it you're doing up there?" I called out. She was quite a sight balancing on the ridge with a roll of tarpaper on her shoulder.

"Fixin' a leak. Hi Jim, welcome to the Garden of Eden." She slipped down the roof and found the ladder with one foot. "Have a good trip?"

"Sure, no problem. Mort made it nonstop all the way just like the book said." No way was I going to tell her I had almost destroyed her airplane through sheer stupidity.

"Where's Richard?"

"He's visiting Gayle. They seem to be hitting it off."

"Come on in Jim, the coffee's hot."

Jim grabbed me by the arm pointing, "Will you look at that." Across the creek in the quakies, we counted twelve elk. "I know I'm going to like it here. When does hunting season open?"

Mary and I would much rather watch the animals than kill them. She changed the subject. "We got a letter from Clay. He says there may be trouble at the Sierra Madre if we don't do something soon."

Dammit, cannot get even a little break, constant pressure from all sides. I said, "Any good news?"

"Yes. The electrician finished the wiring. We can run the grinder now."

"That will help."

"And Norman left some black sands from a mine in Fairplay to test."

"I want to see that. Show me where it is, Mary," Jim said.

Jim went to look for his microscope. There were many different ore samples for him to decipher and they were all complex. He was in his glory.

On our way back from Riggs, Idaho, we had just passed over Montrose and were following the Gunnison River up the Black Canyon. Mort was running perfectly.

"Well," Jim said, checking his notes, "I'm afraid Norman and his bankers won't like what we're going to tell them."

"They lost the vein on the fault, didn't they, Jim? How can we tell which way it slipped?"

"That will take a lot of time to determine, Mac. It will probably cost more to chase that silver than it's worth."

It would cost a lot of time and money to locate the vein again if it could be found at all. An ore vein can be very strong and easy to follow until it comes to a fault zone. These are zones of quake displacements, caused by the earth's crust slipping up, down, or sidewise. Or maybe the system split and fell away. One copper mine in Arizona had a displacement downward of over two thousand feet. The study of the earth's crust is a science of its own.

"We have a problem with this, don't we?" Richard said. "If we give Norman a negative report, he has a chance to lose the financing."

"Which would present us with another cash flow problem," I added.

"Too bad we can't write a report that would please them. We could extend our stay at Bonanza that way."

"Yeah, why do we always have to be honest? No one else seems to be."

We tied down Mort and wrapped a chain around the prop. The case hardened padlock would at least slow down anyone with the idea of theft. I had not forgotten the stolen airplane factory in Mexico.

"Thanks, Mort. Good job," I said patting him on the nose. I knew he liked that.

We crossed the bridge and started up the hill toward the mill building. "Something's missing. What's gone?" I remarked. There was an empty spot on the upper level.

"The cat and loader were there when we left," offered Jim.

"Funny you don't notice things 'til they're gone isn't it?

"Kinda like a girl friend," added Richard.

"What's happened now, Mary?" I asked getting out of the pickup.

Mary was painting the cabin. "Two lowboy trucks and trailers came and hauled them off yesterday," she said.

"What was the reason?"

"They had the ownership papers. Said that Norman had only borrowed the machinery and they had a job for it."

"Norman told us he owned that stuff."

"Yeah, he told us a lot of things."

"Wonder what else doesn't belong to him," Richard said.

"Oh, well, that's his problem. We have a report to write. Can you type, Mary?" Jim asked.

"Of course, but it would help to have a typewriter," laughed Mary.

"Gayle has one. I'll go up and get it," Richard said.

We wrote an honest opinion on the Idaho Project.

We discovered more in the town of Bonanza than the water pump. There was a Hudson Hornet with weeds growing in the back seat and further up the road a garbage dump. Dumps are wonderful things when you are far from a hardware store and do not have much money to buy parts. We had rescued the compressor from an old refrigerator and had just finished installing it in our lab. The compressor had become a part of our vacuum filtering system.

"You guys really fucked me up!" Norman said storming up the stairs.

"What do you mean?" I asked in total surprise.

"You gave my investors a report that turned them off. They won't give me any more money," he fumed.

"We reported the truth, Norman," Jim said.

"You might have hedged a little."

"We don't do things that way," Richard said.

Mary took Norman by the arm. "Calm down. Let's go have some coffee. There has to be a way to solve this."

After work we filed into the cookhouse. Mary was stirring a big pot of chili beans. "Norman go away mad?" I asked.

"He calmed down and was able to understand our side. I reminded him that we had to make true reports or we wouldn't get any more consulting jobs."

"What's his next move?"

"He said he would contact a guy in Maryland, a doctor who has apparently put money into the mine at Fairplay."

"The black sand from there has a lot of gold in it," Jim said.

"Heard some news on the radio that was almost funny," Mary said.

"What did you hear?" I asked with a mouth full of beans.

"Jimmy Carter is suggesting that gas be rationed."

"Good old Jimmy," Jim said.

"He thinks five gallons a week per vehicle would solve the shortage."

"That would work in our case wouldn't it, Jim?"

"Why not. It only takes five gallons to drive down to the gas station."

"Then we could stay there a week and get our next five gallons and come back," Richard said.

"Now we can't get back to town, we don't have any more gas, right."

Mining camps were more fun than television.

Gayle had a telephone because this year she was the Mayor. "Dr. Piller, my name is Eugene MacDonald. Norman Cochran asked me to give you a call." Gayle was leaning over putting a log in the stove. I wondered if we would ever get warm again.

"Oh, yes. Wait a moment, please, I have to finish another call."

"No problem." As I waited, I could see by Gayle's tight fitting Levis one of the reasons Richard had been spending so much of his time here.

Mark Piller came back on the line. "I'm glad you called. I have some questions."

"Yes, sir."

"Norman tells me you design and build your own machines. Is that right?"

"Yes, we do."

"I think that is amazing. I didn't think anybody did things like that these days." I wondered how he thought things were done. "Norman says you need more money to proceed," the doctor went on.

I was not prepared for this. "Mining costs a lot of money," I answered.

"How much do you need?"

"Ten thousand would carry us for a while."

"Okay, where do I send the check?"

"Villa Grove, General Delivery."

"I'll send it now. As soon as I can get away, I want to come out and see what you are doing. Wish I could do things like that." As I hung up, I thought this guy is either awfully stupid or very rich. Norman was some promoter. I was never able to get money that easily.

"That was quick," Gayle said. "Like a glass of wine?"

"No thanks, Gayle. I've got to get back." I had to leave. It was too cozy and besides Mary would kill me if she knew what I was thinking.

"Have you noticed the thermometer?" Richard asked.

"No, but I noticed that my feet are getting colder."

"As bright as the sun is, it's only fifteen degrees. Amazing it doesn't feel cold."

"That's because of the dry air. The humidity is only about five percent."

"What are you working on, Mac?"

"I'm trying to understand this damn Burkett machine. It's not doing the same job that it did in Tucson."

"You think it has something to do with this high elevation."

"The whole principal here is based on airflow. I'm beginning to believe that the altitude is affecting the efficiency."

"The thin air certainly affects me," Richard said.

"Mary has trouble getting her bread dough to rise and look how long it takes to boil water for coffee. I believe the problem is directly related to the density of the atmosphere."

"You may be on to something important. I'll start making some new calculations and see what design changes might help," Richard said.

Mary came up the stairs. "The meter reader stopped by with a message," she said. "Clay wants you to call him in Arizona. Says it's urgent."

"Crap, now what? That means we'll have to go to Villa Grove."

"I'll go along. I need a couple of things from the little store.

"Damn! Wish we could get a phone up here."

"We're too low in the pecking order. We wait until someone dies first," Mary laughed.

"Might be worth killing for, then," Richard offered.

"Where's Jim? Maybe he wants to go."

"He's studying the chemistry books. Better not bother him," Mary advised.

"Jesus, it is cold!" I said coming out of the phone booth. We should invent a heater for these things."

"That's a thought. They have a few million of them stuck around the country. At fifty cents a piece, we could make some money," Richard quipped. "What's Clay's problem?"

"Kinda weird," I said blowing on my hands. "He heard a rumor that Dick was planning to come up here and get Mortimer. Thought we should know."

"How can he do that? Mort is our airplane. I just mailed a payment to Morales last week," Mary said stamping her feet to keep them warm.

"Maybe we should go on into Salida and check," Richard suggested.

"It's starting to snow. Probably better to go tomorrow in the daylight." Mary responded

"This isn't urgent anyway. Dick couldn't be here yet even if the rumor is true."

"Okay. Let's go home. I'm cold," I said.

"You don't really believe that Dick would mess around with Mort, do you?" Jim asked. It was Sunday morning and we were having a leisurely breakfast.

"Naw, he knows we bought the plane from Morales. If he tried something he'd be in deep shit."

"Why not check it out? You could at least sweep the snow off the wings," Mary said.

"That's right. Those wings weren't designed for a static load from the top."

"I've learned that refrigerators have a different purpose here, too." Mary said.

"What do you mean?" Richard asked.

"I have to use it to keep things from freezing," Mary laughed.

We had convinced ourselves there was no substance to Clay's rumor. The shock I had at the sight of Mortimer crouched on the ramp with his prop spinning threw my world into slow motion.

Jim and Richard just stood in complete disbelief.

I located the airport manager and somehow convinced him to stop the plane from leaving. As I floated toward Mort, I saw two people inside. One of them was Dick. Then I saw red. "What the fuck are you doing in our airplane!" I grabbed him by the throat pulling him out on the tarmac. I remember saying something like, "You son-of-a-bitch," as I punched him in the nose. Then I was on my feet. A big cop had picked me off Dick's chest.

"Calm down, mister." I think he said.

"What do you mean, calm down? This cocksucker is stealing our airplane. Why don't you grab him?"

"We already checked his papers. He has a letter of authorization to repossess the plane."

"A letter from who?"

"The leasing company, signed by Alphonse Morales."

"You're kidding me. There's nothing wrong with our contract. Where's a phone? I'll clarify this right now!" Dick was bigger than I was, but he was cowering under Mort's wing. I turned to the manager and said, "How did they get the chain off the prop?"

"I loaned them a torch. They cut it off."

Morales was not at home. He may have been, but he did not answer the telephone.

"I couldn't find him," I said coming out of the office. "What are you going to do about this, officer? Are you going to arrest these guys for theft?"

"No. We've already made the decision to let them take the plane. I might consider putting you in jail."

"Me? What in hell for?"

"Disturbing the peace?"

"You gotta be kidding. I'll go to jail, but this asshole goes with me until this gets solved."

He walked away and talked with the other cop. "We've decided to let them leave. They have proper papers."

"We have papers, too. Why don't you look at them."

He waved toward Dick and said, "Go ahead and leave. I'll handle this."

Both Richard and Jim had not said a word. We could only

dumbly watch as Mort lifted off and disappeared.

"Now what are we supposed to do, Mr. Policeman?"

"If the plane is yours, get an attorney and get it back."

"Did it ever occur to you that once a plane is off the ground there's no way to know where it's going?"

"Oh, that's right. I hadn't thought of that." They could care less about us. Dick had paid off the cops. Small towns are hard to deal with sometimes.

"I had a feeling, should have gone with you," Mary said.

"MacDonald wasn't too diplomatic," Jim chuckled.

"We have to follow them, Mary. Want to go to Tucson? He'll probably go there since he lives there."

"Let's see if we can get a flight south from Alamosa. I think they have a commuter."

"Life gets tedious don't it," Richard commented.

"I'm getting tired of it. What else can go wrong?"

Mary flew from Alamosa and I followed with the pickup. We were at Clay's house. "What are you doing?" I asked. She was sitting on the bed with her back to me.

"I learned that Morales has an apartment downtown, she replied. "I'm going to pay him a visit." She was loading her Smith & Wesson. A couple of days earlier I would have gone with her. One of Bernie's elephant guns would have fit the occasion. I had since calmed down a little and was able to convince her that a court procedure would be better. The way this finally worked out, she was right. We should have killed Alphonse Morales.

"What did the attorney say?"

"This is an open and shut case, a felony theft, since there was no proof of lack of payment. Even if we had been late, they had to notify me by certified letter. He thinks it's a shoo-in. Took the case on a contingency."

"I still don't understand how Dick thought he could get away with it," I said.

"The feedback I got is Morales wanted the plane for running

dope. He hired Dick to go pick it up," Clay said.

"That damn guy is something else."

"Yeah, he's been promoting Burkett's garbage machine. He has taken in a lot of money from investors. Dick just finished building a big unit for him."

"How's the project coming along at the Sierra Madre, Clay?" I asked.

"We got stalled last week. The workers want their paychecks. The machinery is almost ready and a couple of small miners said they would supply us with ore. We won't have the immediate expense of running our own mine 'til we get the mill into full operation."

"That sounds promising. According to the attorney, we'll have Mort back in a week or two. Things are moving in Colorado. We'll be able to help soon."

"Tequila?" asked Clay waving the bottle of Cuervo.

"A double!" Mary answered.

On my return to Colorado, I was busy in the lab. We had just finished part of the test run on a sample of concentrated gold ore. On the bench were four dishes of coarse gold. I heard a voice. "Jeez, would you look at that. Is it all gold?" the propane truck driver and his helper had slipped in the back door. "Naw, can't be!" One of them answered. "There's too much of it."

Security in this situation is very important. I said, "He's right that isn't gold. That is what you call 'fool's gold', a sulfide."

"I knew it," the driver said. "Nobody around here has that much gold. Sign for the delivery would you, Mac?"

"Okay. Here you are. See you next trip." More than six thousand dollars was setting there on the bench. We did not need a truck driver telling stories in the bar about our operation.

I was in the cook shack with a cup of coffee. Richard came in with a letter. "This is from Dr. Piller. Maybe it's the check," he said. The letter contained a personal check for ten thousand dollars.

"I've been talking to Norman about the contract on this property," Richard said. "He says he'll sign it over to us. Their company has decided not to continue."

"That's great. Does that include their down payment, too?"

"Everything. They are taking a tax write off on the whole project. Norman said he had arranged with the owner to have a meeting at his office in Kansas. We can make the changes and finalize every thing in one trip."

"How can we get there without Mortimer." I said.

"Why don't you charter a plane?"

"That's a good idea. Save us a couple of days driving time. I'll go down to Salida and reserve a plane for next week. Want to come along?"

"No thanks. I've got some drawings to finish."

"Where did Jim and Mary go?"

"Up Little Kerber Creek. They're getting samples from another dump."

"I'll go down to the airport. Be back by supper time." I left the check on the table.

GEORGIA

It was Friday afternoon. Things had finally started working in our favor. We had received the check from Dr. Piller that would allow us to continue work on the project and the property was available for our purchase. As I turned onto Airport Road getting Mort back was our only problem and I was feeling good.

I had just completed scheduling the flight to Kansas. "Are you Eugene MacDonald?" I turned and looked up at the county sheriff. Good I thought, now I can tell him our side of the airplane story. "Yes, sir. I'm MacDonald."

"Would you come down to the office, please. We have some things to discuss," he said.

"Sure. I'm finished here. I'll follow you to town."

"Can you ride with me? My deputy will bring your pickup." I guessed he wanted to save a little time discussing what had happened with Mortimer. On the way I tried to start things off by bringing up my thoughts on how the city police had handled the situation. He did not seemed interested in what I was saying.

I followed him into his office. I could see three cells in the back. "Step over here," he instructed. "Give me your belt and empty your pockets."

"What for?"

"I have a federal warrant here for your arrest. You are my prisoner."

"What warrant? What are you talking about?"

"According to this," he answered holding up a document, "you are a pretty bad guy. We were getting ready to pick you up at Bonanza. You saved us a trip."

"What in the hell is this about? I can't believe what you're telling me."

"I can't tell you what it's about, MacDonald. They don't tell me those things." I was in total shock. He had me by the arm and was saying, "Step inside, please." The cell door clanged shut.

"What am I supposed to do?" I stammered.

"Just sit down and wait. I've called the Marshall's office. They'll be coming from Denver to pick you up. Ask them what the problem is. This is out of my jurisdiction."

I sat down on the steel bed and stared at the wall. My brain had stopped working. The world seemed to be spinning. I am supposed to have a phone call, but who would I call? I do not dare try Bonanza. With the party line, everyone will know about this. I know Norman is out of town. "Well, MacDonald," I said aloud, "this is a different kind of dilemma."

"What did you say?" the sheriff called from his desk.

"Nothing, never mind." There was a paperback book on the steel lounge. *The History of the Calico Mine* the title read. I was deep in the Calico Mountains mining for silver when I heard a voice from the outside. "MacDonald, I'm agent Black. We're here to pick you up."

"Glad to meet you. Would you tell me what this is all about?"

He started with, "You have the right to remain…"

"Wait a minute. I want to know why I'm behind bars."

"Procedure is first to read you your rights," he finished the ritual.

"Are you allowed to explain why I'm here?"

"You have been indicted by the Federal District Court in Atlanta, Georgia. A warrant for your immediate arrest was issued."

"What did I do to get all this attention?"

"This document says you are wanted for mail fraud, aiding and abetting and grand theft of more than a million dollars."

"A what! A million dollars!"

"You must have done well with a scam like that, MacDonald." I was guilty in this guy's eyes already. "You are a federal prisoner," he said as he snapped the handcuffs. "You will do as I say."

"Where are we going?"

"Denver. The county jail."

"Sheriff, could you get word to Bonanza for me?"

"See what I can do," he answered

It was a long ride to Denver.

The fellow behind the counter said, "Sign here." It was my personal property list. Wallet, comb, pocketknife, belt, twenty dollars and eighty three cents cash, it read. I sighed and signed. I was lead through a series of barred doors and finally up a steel stairway to a cell on the second floor. "You're in cell block D23E," the guy said. I saw a stainless steel bunk and a steel toilet with no lid. The automatic door slid shut.

Fuck, I thought.

I was seated at a long steel table in the center of the cement floor. I had already calculated how many people were in this cellblock. Now I was counting the cement blocks in the walls.

"Could I talk with you?" a voice said.

"Sure, why not?"

"My name is Chuck. I've been watching you." Chuck was the first of my companions who had offered a word. "You haven't been here before have you?"

"No, I don't understand any of this."

"I figured that. You don't look like you belong here. You should be careful."

"What do you mean?"

"The guys figure you are a ringer."

"A ringer?"

"Yeah, someone under cover, maybe pick up info. A rat." He was not talking very loud. "I don't think you are. Want some advice?"

"Sure."

"Keep your eyes and ears open and don't talk with anyone. There are a lot of tough guys in here." Nice to find a friend, I thought, as he moved away.

The following Thursday, I found myself staring up at a judge."That's it, your honor. I figure this is some kind of mistake."

They had brought me to the federal building in chains.

"Well, you could fight extradition, but that will take some time. I suggest you go on down to Atlanta and get this problem solved."

"Thanks for the advice, your honor."

"Next case," he said.

Eight days later, I was escorted to the airport in handcuffs. The humiliation and embarrassment of being escorted through airports and trying not to notice the stares of the other travelers was excruciating. The two Marshals had little to say on the flight from Denver. I asked what the trip was costing the government. They did not answer.

We arrived at the Fulton County Jail. It was Friday afternoon. Two weeks ago I had started the short trip to Salida. It seemed like years. My protectors saw me to a holding cell and did not even say goodbye.

While standing in the cell, I spotted a pay telephone. "Say fellow. Could I borrow a dime?"

"You owe me," he said.

I made a collect call to Oregon. My brother answered. "Listen, Tony, I've only a got a second. I'm in Atlanta in the Fulton County jailhouse. I'm afraid of getting lost. No one knows where I am. Try to get word to Colorado. Okay?"

"Okay." To my surprise the dime rattled back into the change slot.

"Here's your dime fella. Hope you don't want interest on the loan."

This time the police finished my Sign. It had numbers, my picture and my fingerprints.

It appeared to me that all the inmates in this cell with me had been here before. They knew the ropes and seemed to know each other. "You're a federal prisoner, aren't you?" The guy with the dime said.

"Yeah, how did you know?"

"You don't look like the rest of us. What's your name?"

"Mac, and you?"

"Willy," he said puffing up his chest.

He was a little shit, about five feet four. "What are you in for?"

"I'm a professional bank robber." He stressed 'professional'. "They have me here on an assault charge."

"What happened, Willy?"

"I found my ex-wife's boyfriend had my kids in the fields picking tomatoes. Nobody treats my kids like that. I shot his knee caps with a shot gun." This is the kind of guy I always wanted for a friend. "What do you do?"

"I'm a miner."

"You mean you dig tunnels and stuff like that."

"Sure, that's my job."

He looked thoughtful as he wandered off. A news article on the TV was showing a picture of a plane. It looked like a Lockheed Loadstar, and a shot of the pilot. The reporter was saying they had apprehended this fellow with a full load of Columbian marijuana. It was Saturday night. The guard had informed me this was only a holding cell until they could arrange permanent facilities. The bars opened and a new guy came through. It was the pilot on the TV show. I'll be damned.

He sat down on a steel bed close to me. I said, "That Loadstar is a nice airplane, isn't it."

"You know about planes?

"Yeah, flown some. What happened?"

"Aw, I made a damn fool mistake. Thought there was fuel where I first landed and when I found there wasn't, I just took off. The tower thought it looked suspicious and had me followed."

He got up and stretched. "I'm tired. Flew a long way today." He did not seem anxious or nervous. When I woke up, he was gone. Someone had paid his bail in the middle of the night. I wished I had connections like that.

Time forgets to move when you are locked into small spaces. Two or three days can be an eternity. Because of my ignorance of how the rules of federal justice worked, I had no reason to be afraid. Being alone with my thoughts was no problem. After all, that was a way of life for me. The problem was being amongst a group of alien thinkers. I did not understand their world.

"Hey Mac. I've been working out a plan." It was Willy. "I want to see what you think."

"What's up, Willy?"

"Would you like a smoke?"

"You bet. Thanks." He handed me a package of Top Tobacco and some cigarette papers.

"You say you could dig tunnels, right?"

"Sure."

"I know of a bank that always has at least a million dollars in it. I've always worked alone, never had a partner, but this job has to have a tunnel." After living a few days in my new world, a discussion like this seemed normal.

"What's the plan?" I said rolling a cigarette. My pipe had been confiscated in Denver.

"My brother was the electrician when they built this bank. I have all the drawings of the security system."

"How long would the tunnel be?"

"Two hundred fifty feet or so. You dig the tunnel. I'll take care of the rest. We split 50-50."

"Hmm, this is interesting." I was actually trying to think of a way to help him. "Where would you start digging, Willy."

"There's a vacant lot across the street. You would have to work at night I suppose." He has to be kidding I thought. He was getting more excited as he talked.

Willy was not kidding. "There's a lot of planning to do. We can work it out at State."

The Georgia State Penitentiary was second home to these guys. Willy assumed I would enroll at State with him. He had not graduated from there. He was on sabbatical, learning how to do banks more efficiently. "My wife will bring us pies and cakes on Sundays," he went on. "A year will go by pretty fast. I did a nickel last time." A nickel was a five-year stretch.

"Let me think about it, Willy. Give me a day or so."

"No. They are moving me tonight. I need to know soon."

Later when I told him I did not want to join in his new venture, I realized the error in talking with him at all. The guy in Denver had warned me about it. I had become an enemy because I knew his plan. This time I was lucky. Willy was assigned to a separate section and I never saw him again. He could have killed me.

These diversions were helpful. Without them I might have gone totally bonkers. I still did not understand why I was here.

It was Monday morning. Seventeen days and there had been no contact with anyone who could explain one single part of this fiasco.

"Get your stuff together, MacDonald. We have prepared another spot for you." The guard addressed me.

"What stuff? I don't got no stuff." My vocabulary was improving. All I had to pack was the one book that I had pinched from Salida.

"You should like this new accommodation." We walked through a labyrinth of corridors and cells. "Enjoy." It was becoming obvious they did not like holding federal prisoners. Even the door slammed shut with what sounded like vengeance.

Christ, what is this? It is a goddamn zoo. I was looking at a scene from the Planet of the Apes. A tall colored fellow motioned me toward a mini-cell. He was not just colored. He was blacker than anyone I had ever seen and completely bald. "I'm Frank. I run this place. What's your name?" I was not sure who I was but I told him my name. "Okay, Mac. You sleep there."

This was a real jail. There were four separate cells within the cell. Each had a sliding door that opened and closed automatically. I counted eight double bunks and fourteen animals, plus one white haired old fellow in a corner reading a book. In more than two weeks, this was the first book I had seen other than the one that I had borrowed from that cell in Salida.

Something else was upsetting my nervous system. What was wrong? It was noise, total chaotic noise, each of these monkeys had a transistor radio and every one was playing hard rock on fourteen different stations. "Jesus, this will be impossible," I said aloud.

"What," the white haired fellow said.

"The damn noise."

"Oh that. You'll get used to it. I'm George, George Clark," holding out his hand.

"Mac." He spoke in a language I could understand, American English. I lay down on the steel and tried to adjust myself to a world I never knew existed. The minutes were sluggish, and dull and flat.

Now it was Thursday. I found myself in a sort of Hearing Room. My ears were still ringing from the jungle atmosphere of my new home. The girl was talking to the guy with a robe while someone was fiddling with a recording machine. The girl reminded me of Mickey Mantle. She turned and said, "My name is Glenna Stone. I am the Assistant District Attorney." She put a lot of stress on 'I'. "This is an arraignment and a hearing to set a bond. Your honor," she went on, "the government wants a bond to be set at five hundred thousand dollars."

I did not know what bonds were about. "Where am I going to get that much?" I blurted.

"The law only requires a ten percent deposit, Mr. MacDonald," the judge said looking amused.

"Only? That's fifty grand!"

"Your honor," my friend Glenna said, "this man has taken over one million dollars from the unsuspecting public. This should be no problem for him. What's more, he has been living in a condo in the Colorado Rockies with a kept woman."

Your honor, I don't know what she is talking about. I know nothing about a million dollars. I don't have fifty cents." I had forgotten about the money they had taken from me in Denver.

"Do you have an attorney?" the judge asked.

"No, I can't afford an attorney."

"We'll get you one. Glenna, we'll set this up for three o'clock tomorrow."

I went back to my surreal existence.

A public defender was assigned my case. His name was George Washington. He was as black as Frank, my new cellmate, and he had hair. "From what you are telling me I believe we can get the bond reduced."

"If they reduced it to zero, I'd have trouble raising the money."

"Glenna will never believe that. She thinks you have a Swiss bank account."

"Where in the hell does Glenna get all of this high class information?"

"From the witnesses at the Grand Jury hearing."

"What Grand Jury? What witnesses?"

"You didn't know about that?"

"How would I know?"

"Glenna's star witness was an Alphonse Morales. He told them the whole scheme was planned by you."

This bit of information was hard to assimilate. While my brain was rapidly searching for a spot to store it, I said, "Interesting! What happened to Goldstein?"

"I believe they have him in custody." How in the hell did Morales get… "How much money can you raise for a bond?" George was asking. He seemed far away.

"Nothing. Nothing at all, we don't have anything."

"We can bargain with Glenna, but I can't ask for release with zero bond. Could you come up with a thousand?"

"I don't know. Can you contact Mary Martin for me? She'll think of something."

I was to learn later that this was the first federal case that George had handled. He had just graduated from law school.

While waiting in the courthouse holding cell, I met an inmate who had been involved in a big gold heist from the airport in Chicago some years before. He gave me the name and number of his attorney.

George was saying, "I was able to talk to Miss Martin. Glenna wants your passport, too."

"Why the passport?"

"She's afraid you'll skip to Switzerland," he laughed.

"George, I can't handle much more of this lock-up business. I'm getting a little bit crazy." I could tell he did not care.

"Miss Martin will be mailing these things to me. I'll contact you when they arrive."

"Mail! Do you know how long it takes to get mail from Colorado?" These people should be made to spend some time in here. Every minute is torture. "Why not Fed Ex or UPS? There are better ways than mail!" I almost screamed.

George was not listening. He had other things to do.

Each day we monkeys were released from our cages for an hour of exercise. I did not go with them. One whole hour of quiet was something to look forward to and to really appreciate. During this hour I could reorganize my system and think more clearly. The Morales news had been disturbing. The pieces began to fit. Morales had known about my indictment. That was why he made a move on Mortimer. He must have figured that I had been picked up by the feds and could not cause any problem. However, Dick had been too quick. He went after the plane before they grabbed me. It was not difficult to build a strong feeling of hatred toward these idiots. I could picture Mary loading her revolver. Quit it MacDonald. This is not good. Get your mind on something else. The menagerie was filing back. The noise began its crescendo.

I was watching one of my bedfellows braiding another's hair. They took turns at this. It was very educational. "Say Reginald, what you in here for?" one said.

"I'm AWOL from the navy."

"Why'd you run off?"

"I'm afraid of water," Said Reginald.

"Man, didn't ya'all know that ships use water."

"Yeah, Reggie, you seen it on TV," another piped up.

"Sho man, but I didn't believe it."

"Hey honkie, wanna play me in chess?" Frank the pit boss asked.

"Sure, man, I'll play with you." I learned another lesson. Do not beat the Champion of the Apes, especially when one is confined in his forest. The king had not lost a game in six months. His image was tarnished. I became a threat to his domain. Under these conditions, one should not sleep too soundly at night.

George Clark was interesting. "George, you don't seem anxious about all this. How do you take it?"

"I've been here before. The first time, I was about fifteen years old." He stared into the past. "This place was a lot different then. The latrine was a trench running in the floor through the cells. At night we slept on a long bench all chained together, twenty to twenty five of us."

"Shit house! That must have been real comfortable."

"Yeah, it was. If someone had to hit the can in the middle of the night, everyone had to go with him."

"How long do you have to stay this time?"

"Forever. They have me for life now. I'm what you might call retired."

"Life? What did you do?"

"Nothing much, at least I didn't hurt anybody. I'm a counterfeiter by trade."

"I've never met a professional counterfeiter before, George. How did you get into that field."

"When I was twenty, I was sent to State for a five year stretch. I went to school there and learned to be a printer. I decided to join 'society' when they turned me loose, but I couldn't make much money from printing business cards and letterheads. I came up with a plan that worked for the next forty years. I've made a good living. Now I'm going to sue the State."

"How can you sue the State of Georgia? For what!"

"They stole ten thousand dollars from me! I want it back."

"What happened?"

"I developed a good business, Mac. The first thing I did was to build a portable printing shop. Everything would fit in my car and a little trailer. I would locate a town that had a large payroll like Lockheed, then rent a small shop and advertise for work. I opened an account at the bank that had the most branches. I usually started with ten thousand dollars on deposit. I printed up my own driver's license, birth certificate, Social Security card and the like. The whole thing was perfectly legitimate."

"Except for creating a false person."

"That didn't hurt anybody though. It was fun coming up with new names. One time I was George Herman Ruth. Nobody noticed."

"How did you make money? Print it?"

"No, no," he laughed. "I'd get hold of a payroll check from the major company in town and print checks made to me, my current name. Let's say there were twenty five branches of the bank around the city I'd chosen, okay?"

"Okay."

"I made up a map of the best and quickest connecting route between the branches. I would print checks made out to me in my current name. The checks were made out for odd amounts, like two hundred nineteen dollars and eighty three cents, always within the general weekly amount the average worker was making."

"How long did you usually stay in town?" I was getting interested in his method.

"From four to six months, never more than a year. Depended on how large the city was. I could spend a year in Atlanta."

"How did you get your original deposit back?"

"After I got rolling I would deposit the money in the main branch and get a passbook. I used the initial deposit for my operating expenses. I withdrew a little at a time. I planned the final move about the time my original balance was used up.

"I could pick up from two hundred fifty to three hundred in cash by printing cards and letterheads and circulars for the local people. When I decided it was time to move, I put my shop back in the trailer and packed the car. I always moved on a Friday, payday. At ten o'clock I hit the first branch on my escape route. Using the passbook I took back cash to an even dollar number and would deposit the smaller amount, like twenty-one dollars and fifteen cents. I could pick up two fifty, three hundred in cash at each bank. As the years went by the wages went up so I could print bigger checks."

"How much could you make?"

"Last few years I averaged about forty grand a year. Would have made fifty this year if they hadn't caught me! That damn old woman."

"What old woman?"

"The one that stole my ten thousand. I'm going to sue, I tell you. It wasn't fair!"

"Well, how did you get caught after so many years?"

"I'd had it too easy. I got careless. Usually the banks had a young girl taking new accounts, public relations, you know. This time I got an old girl that had been around for several years. Normally, I would have rented a building and moved in before the account was opened. I'd looked at a building that morning and gave her the address. But I didn't rent that building. I deposited my ten thousand and went back to my hotel. For some reason she got suspicious. Maybe something I'd said. I don't know. Anyway she sent out an investigator to check and he found an empty building.

"The officials had been trying to find me for years. This old girl put two and two together and figured I might be the guy. So, here I am. Retired."

"But how do you have a case to sue?"

"They confiscated my deposit. That was my money, damn it. I earned it."

The State had done a good job. They had taught him a useful trade.

I felt each day grow longer and longer. George lent me a couple of his paperbacks, Biography of Willy Sutton, the professional safe cracker, and Butch Cassidy and the Sundance Kid—textbooks for George I guess.

It was Friday when George Washington had me paged in my office. "I got the stuff from Colorado and cleared you with Glenna. Miss Martin sent a credit card and a note," he said handing me an envelope. "Pick up your stuff at the properties office. You're free to go."

I was not prepared for this. Things had happened too quickly. Here it was late in the afternoon on a Friday. It had been five weeks since I had made the short trip to Salida. Now I was standing on a sidewalk outside the bastille in Atlanta. " What the hell, MacDonald. You haven't the foggiest idea where you are in this city or where to go," I was still talking aloud. "One thing I do know though, it's time to get out of Dodge."

The property clerk had given me a county check in replacement of my twenty dollars. Across the street I saw a bail bondsman's office. Maybe they will cash the check.

"Are you kidding me? We wouldn't touch a Fulton County check with a ten foot pole." That was helpful. Back on the street, I spotted a yellow cab.

"Hi there, could I talk to you a minute," I asked the driver.

"What do you need?"

"What's the fare to the airport?"

"Sixteen bucks."

"Let me make you a deal. I've got a county check here for twenty bucks." I showed him the check. "I'll sign the whole thing over to you if you'll take me."

"Looks real. Okay, endorse it to Joe Brown. Got any bags?"

"Nope, just me."

"Let's go. You can ride up front."

It was quite a ways to the airport. I kept glancing back to see if anyone was after me. Maybe they had changed their mind. "What do you think of this Jimmy Carter as president, Joe? He was from here wasn't he?"

"No one around here can figure it out. When he was Governor, he never seemed to do any thing. How he can run the United States is anybody's guess."

"Politics has its weird ways."

"If the South had won, things would have been different," Joe, the Reb, said.

I was not sure what I had expected to do at the airport. But there was a sense of urgency surrounding me. Glancing at the schedules, I spotted a flight to Denver, time of departure twenty minutes. The flight number was flashing 'LOADING'. Rushing to a ticket window, I tossed the girl the credit card. "I have to catch the flight to Denver. Can I make it?"

"Yes, sir, but you will have to run." She did not notice the name of Mary Martin on the card. Not until the DC 8 was moving away from the gate did I relax. No one was chasing me. I was actually free. What a feeling!

Life gets interesting when you are out in the world with no money and no friends. I had made a collect call from our stop in St. Louis and left a message with Gayle that I would be at the airport. The wait at the Denver terminal was a short four hours. Here, at least, there was room to walk further than fifteen feet in a straight line. Richard, Mary and Gayle found me in the baggage section.

While driving back to the mill at Bonanza, the group filled me in about trips to Arizona to get machinery and how they had made a contract with the Fairplay people to process the concentrates from their mines. Richard had gone to Kansas and had completed the purchase contract with the owner of the property in Bonanza. They had been busy while I was on vacation.

Bonanza presented another problem. It was too quiet. I could not sleep. I could not get my mind organized. I was still in shock. "Well, Mac, what's the next step?" Mary asked. I had given them a synopsis of my journey into the unknown.

"I think you should get another attorney. Sounds to me like George Washington doesn't have enough experience to handle this," advised Richard

"That's going to cost a lot," I said. "Where can we find money for that?"

"Why not ask Dr. Piller? I've talked with him twice. He may help," Richard said.

"Wouldn't that be dangerous? He could cut off the relationship if he finds out I've been indicted."

"He'll find out sooner or later, anyway. Better to tell him now," Jim said.

"Guess you're right. Have to do something even if it's wrong. So what have you guys got on the burner?"

"We're ready to operate on a small scale, but to do what we would like, the building must be expanded," Richard said. "I think we may be able to get a bigger contract from Fairplay to process their concentrate. The agent for the Company is Barney Spielman. He said they could advance funds to expand," Jim explained. "Also my son, Bob, is coming over. He can help with some of the work while you get your problem solved."

"Sounds good! What happened in Tucson, Mary?"

"Bad news. The judge wouldn't release Mortimer. The case is getting more and more complicated. According to the attorney, Dick was supposed to deliver the plane to Morales. He didn't. He is claiming it is his."

"Why can't we get Mort? The old boy belongs to us."

"Morales pulled a slick one. He sued Dick for theft. The court will hear that case first then make a new trial for our side."

"That fucking Morales oughta have his balls cut off," Jim offered.

"I'm sure he doesn't have any," Mary replied.

"How are you coming with Monty's system, Jim?"

"I have it well developed. Bob can help me get it into production. There is a lot of welding to do on the tanks and piping. I'll get back to work now. Come on, Richard. See what you think of my new idea."

They were gone. Mary said, "Mac there's something wrong with Richard. He is not acting his usual self."

"What do you think is the problem?"

"He's spending more time alone in his camper and doesn't agree with some of the things Jim has him working on."

"Two operating engineers on the same job is tough. It will work its self out."

"Another thing, Richard is not seeing Gayle anymore. He won't talk to me about what happened."

"I think she drinks too much and maybe he's worried about the problem with Mrs. Brown."

"I thought he had that covered."

"I don't think so. Something wrong there. On the new attorney idea, I heard of one in Buena Vista that takes these kinds of cases," said Mary.

"Who is he? I'll go see him."

"Sounds very complicated. Working a federal case like this one will take a lot of time," the lawyer said.

"Will you help me on this?'

"Sure. I'll need a retainer to start, though."

"Good, I feel better already. How much of a retainer?"

"Twenty thousand to start and five thousand a month, probably total cost hmm-hundred thousand."

"A hundred thousand dollars!"

"That's all, plus expenses, of course."

I did not feel so good. "Okay, I'll let you know." You son-of-a-bitch. I was more mad at myself than him. Thanks for your time," I said as I tipped my hat and walked out the door.

Dr. Piller was more understanding. He loaned me ten thousand for my legal fund. I found the attorney in Atlanta who had been recommended by my friend the bullion thief.

"I like this case of yours," he was saying. I had arrived in Atlanta the night before. "How much money can you raise?"

"Ten thousand," I answered.

"Well," he said, rubbing his palms together, "that's the exact figure I had in mind. I won't be able to handle this personally. You will work with my associate, Gregory Brown."

"Is he white?"

"Yeah, he's one of us. Give me the money, we have to get cracking."

I handed him the ten thousand in cash. This decision to get our own attorney was a major error. I had convinced Glenna, the Assistant DA, there was no money. Staying with George Washington would have made things much easier and cheaper. I earned another Sign.

"Mac, I think we can get this indictment over-turned. You're no crook, from what I can see Glenna has no case," Greg said.

"Somebody decided I am. What happened to the great thing of 'innocent until proven guilty'? Where is Glenna coming from?"

"This is her first case, Mac. She has a large budget at her disposal to stop white-collar crime. If she can get convictions in this case, she will be able to start her own law office and draw in many clients. She's after the notoriety."

"So she can feather her nest at the taxpayers' expense. Is that it?"

"That's how it works. I don't like it. Wish I could do what you do, live out in the woods. I've arranged a meeting with her at two. Let's have lunch."

"I was inclined to believe in Mr. MacDonald's non-complicity in this scheme," Glenna was saying. "But something new has turned up to convince me to go ahead." She had a smirk on her face.

"What new evidence do you have, Glenna?" asked Greg.

"Greg, I want you to listen to this!" Strange that they are on a first name basis. Glenna pushed the start button on a recorder and we listened to the meeting that had been held in the U.S. Gold office in Atlanta. Where could she have gotten this? "So there you have it," Glenna said when the tape ended. "This places Mr. MacDonald in a board meeting with Danny Goldstein, Paul Levine and Mr. Samuelson. Obviously he knew about the whole scheme."

"That was no board meeting. Who gave you the tape?" I asked.

"Danny Goldstein. He told me you were the one that had planned everything."

"Come on Mac. Looks like we have to regroup."

On the courthouse steps, Greg said, "I'll go find out about this development, Mac. Call you in the morning. There's a nigger in the woodpile." I ordered a double Jack Daniels and two beers at the first bar I could find.

"We're in deep shit now, Mac. Glenna's going after you. There will be a trial."

"But why? Why won't she let me discuss that tape?"

"I've already explained that. She has made a deal with Harold and Danny. They apparently made an agreement with her to implicate you."

"But if she already has them, why me, too?"

"She wants to flex her muscles, Mac. She's still trying to locate Roger."

"She's spending a lot of money on this, Greg."

"Tell me about it."

There was nothing to do but go back to Bonanza, work and wait. We built an addition on the mill and dug a trench with the backhoe for a water line from the creek.

"We'll dig this five feet deep and fill it with sawdust," I told Richard.

"That should be plenty," he said. We were wrong. The line would freeze solid and not thaw for six months, so much for planning and engineering. Extreme cold was one thing we had never dealt with, a good case for learning by doing. Jim had the chemical plant about finished. We were maybe three weeks away from starting production.

Mary went with me to Atlanta for the trial. A little research on living facilities lead us to the hotel across from Loew's Grand Theatre on Peachtree Street where the premier of Gone With The Wind had been shown. We were rubbing elbows with the famous. The Georgian Terrace was where Clark Gable and Vivian Leigh had stayed. We were able to acquire a suite of rooms at a reasonable monthly rate. At least we could live in style while we had our battle with Glenna.

We entered the courthouse. There was a long line of people. My mind was on finding Greg. Mary took me by the arm and said, "That looks like Clay." Glancing at the group, I spotted the Bishop, Jim Stokes.

"What the hell is this? What are you doing here?" Mary pointed to Bob Slagle as I started toward the Bishop. "Jim! What are you doing here."

"Hi Mac," the Bishop said shaking hands.

"Why are you here?" I repeated.

He gave me a big grin. "I've been subpoenaed as a witness. Everyone here is a witness," he said. Then added, "for the government."

"You're a witness for the government? Where are you staying?"

"Don't have a place yet, just arrived."

I gave him our hotel telephone number. "Call us. We should talk. I saw Clay going in a door. Give the number to Clay, too."

"Mac, we were instructed not to talk with you or anyone connected to the case. I'd better go. I'll call you later."

"What are you doing down here," Greg said. "I've been looking for you."

"Came in the wrong door I guess. Why so many witnesses?"

"Glenna is trying her best to impress her boss. She has brought sixty-five people to testify. I didn't get a chance to tell you. This is a joint trial. She has indicted Paul Levine and plans to try you both at the same time."

"I don't have any connection to Levine. How can she do that?"

"Glenna can do anything she wants. She thinks she can prove you never intended to honor your contract to U.S. Gold. She is trying to prove you did not have the credentials or knowledge to supply any gold. She's on a roll now especially since she has the Goldstein boys in jail."

"They're in jail?"

"Both Harold and Danny. They waived rights to trial and pleaded guilty. Glenna gave them lesser sentences for that."

"Why?"

"She made a deal with them to help her get you and Levine. They signed affidavits that you and Paul set up the whole thing."

"That's ridiculous," said Mary. "This whole thing is nothing but a circus."

"A one ring circus. Glenna has made herself the center of attraction," Greg said.

"Who else has she brought in?"

"Her star after Goldstein, Alphonse Morales."

"Morales!!??"

"Yeah. He's a real charmer."

"Rasputin," Mary said. Cocksucker I thought.

"What happened to you guys? We expected you earlier," Mary said. Clay and the Bishop had arrived together.

"What a fiasco this is. That girl Glenna is totally disorganized. She kept us there two hours longer than necessary," Jim said.

"She has every one of the people she subpoenaed mad at her. We stood out in the hallway most of the day while she verified who was who," Clay said.

"Why are you her witnesses? Aren't you supposed to be on our side?" Mary asked.

"I tried to tell Glenna that the first time she contacted me. She would not listen. She insisted on having me here," the Bishop said.

"Who else is here we know?"

"I saw Bob Slagle and your friend Dick Nichols. Dick came over on the same flight I did," Clay said.

"Nichols is here? After stealing the plane how does he have the nerve?"

"Morales is real buddy-buddy with Dick. He has appointed himself head honcho among the witnesses," the Bishop pointed out.

"I don't believe this! What's with Slagle?"

"He's mad because you didn't finish the contract on the Alaska deal. He's out to get you," said Clay.

"How could you afford to fly here?"

"Glenna paid the way. Everyone has a round trip ticket, plus we are being paid per-diem for our living expenses," Jim said.

"She did this for everybody?"

"Every last person," said Clay.

"Who pays for all of this?" Mary asked.

"The taxpayer, all of us," said the Bishop.

"I've got an idea," Mary said. "Why don't you stay here. We have three bedrooms."

"That would be fine by me," Clay said.

"We're not supposed to associate with the enemy, but it would be an advantage. I'll sign my per diem check over to you. Help you with the expenses," The Bishop said. "How about you, Clay?"

"Sure. Have to stay somewhere."

"Glenna is so messed up she'll never catch on. We'll be able to compare notes easier on each day's proceedings," said the Bishop.

"I'm tired, where's my room?" Clay asked.

I was in the hallway while Glenna was trying to prepare for the jury selection. "Are you MacDonald?" a nice looking girl was addressing me. "I'm a reporter for the Post. I've been assigned to do an article on this case. Can we arrange an interview?"

"You bet we can. I want to get this out to the public. Someone has to know how crazy this whole thing is."

"Where can we meet?"

"The Georgian Terrace Hotel. Call us tonight."

Greg approached her and asked, "Who are you?"

"I represent the Post. I've arranged an interview with Mr. MacDonald."

"I'm sorry. This is a closed trial. No reporting allowed."

"Who decided that, Greg?"

"Glenna and I agreed to a gag order on the press."

"Thanks a lot." Wonder whose side he is on anyway. A chance was lost to get out my side of the story. Why did they have a gag order on this?

Somehow, Glenna had stretched the jury selection to three days. Each evening after work we compared notes. "I talked with an interesting witness today," the Bishop reported.

"Who have you found?"

"You won't believe, the Registrar from the University of California."

"The Registrar? Why?"

"She doesn't really know herself. She thinks Glenna wants to show that you were not a very good student. The Registrar doesn't know you from Adam."

"I graduated! Doesn't that count?"

"Glenna wants her to show that you got a 'D' in Biology."

"What is this? Doesn't she know the difference between Biology and Geology? Besides I took that course over and got a 'B'."

"Seems like she is picking at straws," Mary said.

"The total exercise is stupid. Some of the witnesses are sophisticated investors. They feel Glenna forced them into this. The time involved away from their work is costing them more than they lost in the first place."

"Have you found anyone that lost a lot of money?" Mary asked.

"No! None of those people would give Glenna the time of day. They had already written if off as a loss and could not waste the time. They had figured it as a risk investment in the beginning."

"Why does she think you can help?" I asked the Bishop.

"She wants to use me as an example to show that I was cheated."

"Cheated? How did I cheat you?"

"You didn't. I have a surprise for her. That woman is costing me a lot of money by forcing the subpoena on me."

"What kind of surprise?"

"You'll see! I'll be called on the stand first. I have to get to a board meeting in Dallas. Glenna has condescended to let me be her first witness. I think we can put a wrench in her gears from day one!"

"Bishop, you've been in business for a long time. Do you see any reason why the government is putting so much effort toward destroying us?" Mary asked.

"In my opinion I think they have forgotten what the United States is all about. The direction that our politics is taking us tells me the idea now is pushing toward a one-world government. We have become a means to an end. Your family and mine were the pioneers, did all the initial hard work. The last hundred years had brought our country to the major power, controlling most of the world.

"Do you recall the parable of the goat-herder who was given the riches of the world when he picked a flower? Remember he lost everything when he left the flower behind. The flower was the reason the opportunity was given to him. Well, I think that is what is happening to us, the workers, the doers of the world."

"In other words our leaders have forgotten the main reason there is a United States," I said. "They either have forgotten or don't care. I remember one of J. Paul Getty's observations. He said the bankers view people as collateral. The masses are expendable.

"That's right. He made that statement after Mexico appropriated his oil wells," The Bishop said. "He invested the risk capital, did all the work. When he was successful, the Mexican government took over and tossed him out. They knew he was more powerful than they were. So, they had to get rid of him. You are being treated the same here by our leaders, on a smaller scale, of course."

"How can that be?" Mary asked.

"You have spent years learning how to mine and process gold, right?" the Bishop answered.

"Yeah," I responded.

"With your knowledge you can create your own money. The government has no control of you or how you think. By law, they must buy your gold if you want to sell it to them. You can either keep the metal or use it as collateral. What's more, you don't have to deal in US dollars. Any currency in the world is available if you want it. If you could manage to keep enough of your gold, conceivably you could start your own country."

"We wouldn't go that far." I said.

"The government doesn't know that, but you could. The problem they have with people like you is there's no control. They have no way of knowing what you are doing as far as income is concerned or how much bullion you have stored. The bottom line is they can't collect taxes."

"This is the real reason then why public lands are being withdrawn and called wilderness zones," I said.

"Of course! You know that these wilderness zones are where the major mineral deposits are. By withdrawing them and barring commercial entry, they have effectively blocked the miners from developing their own wealth. You had it made, Mac. You had your own mines and the ability to make your own money. Unfortunately the Goldstein deal advertised the fact that you were out there and the government had to find a way to block you."

"Is this why Glenna is so tough?"

"No! Glenna has no idea about this. She is a pawn of the power, too. All she knows is her duty is to put 'bad guys' in jail. Glenna is a product of the new age. She was never taught what we understand. And she'll never know how much damage she is doing to what we knew as the good old USA."

"How do you see our chances of getting clear of the charges she has against us?" Mary asked.

"This case has gone too far. If Mac gets out of this with his skin, he'll be fortunate."

The opening statement made in the courtroom was crushing. Up to this point, I had looked on Glenna Stone as the adversary. My world changed completely when I heard the Bailiff announce; THE UNITED STATES OF AMERICA VS. EUGENE MACDONALD. From here on, the proceedings were like a bad dream.

Glenna was in her glory. She was all puffed up like a bantam rooster. This girl is something else I thought. I wondered if her lady friend was in the gallery. She had gone through the formula to establish that Jim was a sophisticated investor, a credible business man and a Mormon Bishop. As promised Jim was the first witness.

"Now, Bishop Stokes, do you know the accused, Eugene MacDonald?"

"Yes, I do."

"Have you ever had any business dealings with him?"

"Yes, I have."

"Did he ever offer gold to you?"

"Yes, he did."

"Did you pay for this gold?

"Yes, I did."

"Did you ever receive this gold?"

"Yes…" Jim opened his brief case.

Glenna slowed for a moment then asked, "Was it real gold?" What a stupid question I thought.

"See for yourself," he said as he began to stack one-ounce bars of gold on the judge's bench. Bar after bar of bright shiny gold, totaling thousands of dollars. It was beautiful! This was some of the gold he had purchased from Elodia in California. The judge stared at the gleaming bars. The jury leaned forward in their seats. The attention of the entire courtroom was riveted on the stack of gold—real gold. Glenna swallowed deeply and stopped breathing momentarily. The silence was deafening.

Up to now it had never occurred to Glenna that gold was something real. She only knew of it in the form of cheap jewelry. Jim had set her up.

Even more stunning than the sight of the beautiful gold was Glenna's strange reaction. She looked around in a bewildered way, covering her eyes with her palms. Tears were suddenly streaming down her face as she ran from the room. In a Federal Courthouse in front of a jury and a proud judge, the government's star number one draft pick from the highly touted Emory School of Law in the grand State of Georgia, completely broke down.

The courtroom was abuzz. The members of the jury were looking at each other confused. The judge was stunned. It took him several seconds to declare a recess.

Law 101 had not prepared Glenna for the position she had put herself in. "Glenna," the judge called her by her first name. "You have had over a year to prepare this case and you spent a lot of the taxpayer's money, but at this point I am inclined to throw it out of court. What do you have to say?"

Things had taken an interesting turn. Ms. Stone was on trial. Glenna stammered illogically, grasping at straws about the long hours that she had spent and how tired she was. "All right I am going to recess this court until next week. If you don't get your act together by then, we'll stop this whole thing."

The gavel went 'thud' and we went home.

This fiasco took place on the Thursday before Easter. Clay came into the living room of the hotel suite, "Now here's something else you won't believe."

"What now?" Mary asked.

"Glenna has given us all round trip tickets to go home for the Easter holiday."

"What? Tickets for the sixty-five people? Why?"

"She doesn't want to pay the per diem to hang around for five days. She thinks she's going to save money by sending us home."

"Christ, that will cost at least thirty grand."

"It's the shits, ain't it? If only we had thirty thousand."

"Wonders will never cease," Mary said. "Think she would give us a ticket, too?"

"Fat chance," I answered. "Where do they find all these brilliant people anyway?"

"They work for the government," Clay said. "See you next week."

"What is our next step, Greg? Glenna having her nervous breakdown should be a break in our favor."

"We think she will regroup and come back swinging. The judge will allow her to go ahead."

"She has a bunch of witnesses to put on the stand. Seems to me, we're looking at two or three weeks more of this bull shit. Mac, we are getting ready to organize your witnesses. Have you been in contact with any of them?"

"The bankers, attorney, and suppliers, they'll all come. But I don't want to bring in Monty."

"Why not? He's the one who taught you how to run the chemical recovery process."

"Monty is your typical mad scientist. Glenna will make him out as an alchemist of old. I can't and won't chance it. Anyway Jim Scott can cover this part of it."

"Okay, sounds good. We have another matter to discuss."

"What matter?"

"This case has taken more time than we thought. We're going to need more money to proceed."

This was something I had not expected. "I gave you all the money I had, Greg."

"Well, it's all gone. You will have to get us some more."

"Where? How?"

He shrugged and said, "That's your problem. Criminal cases don't make us money. In order to proceed, we will need more."

"Mary, how in the hell can we bring witnesses to off-set dear Glenna's allegations plus pay Greg and his company?"

"I don't know, Mac. I don't know anyone that will even say good morning to me anymore let alone put up cash for something like this."

"Dr. Piller won't advance any more until the mill is running. We can't sell anything. We need everything to operate."

"How much are we talking about, Mac?"

"The guys on the coast will come and testify for no cost but they want us to pay the air fare and food. Maybe about ten thousand for them, and Greg is suggesting they want another ten thousand."

"Impossible, I don't see where we can get it."

"There isn't a bank in the world that would lend us anything under these conditions. Mary, do you remember when the judge in Chicago offered me that deal. He would have let me go if I agreed to get out of the gold mining business?"

"Yes, I do."

"You know what's creeping into my thick skull? All this time we've been looking at this problem as an economic thing. It's not! It's political. The gold dealings in the world are purely political. The economics are secondary."

"I think you're right. It's obvious that the governments don't give a damn about what happens to their people. You know the saying 'too soon oldt - too late schmart'."

"So, now that we've learned a big lesson what can we do about our infinitesimal dilemma?"

"Pour us a drink. We'll think about it. There has to be an answer."

The trial proceeded as Greg predicted. We tried to put pressure on Morales by showing how he was an insider. The part about him walking out with silver bars stuffed in his pockets did not get us any points. Alphonse had said this was a payment for his consulting fees. Goldstein owed him money on his contract. It did not seem to matter that he had the combination to the safe. We were getting closer to our day in court and had no way without money to have our say.

Greg was saying, "If you will enter a plea, we'll finish this up, Mac."

"What kind of plea?"

"If you plea nolo contendere, Glenna will let you go. She's in trouble."

"What does nolo mean?"

"That you are not guilty or innocent."

"But we're innocent."

"No. You're guilty of having no money."

"This is crazy!"

"Our recommendation is you do the plea. It saves everyone time and money."

"To me this is quitting. I'm a fighter, not a quitter."

"Under the circumstances you have nothing left to fight with," Greg advised.

"I'll think about it," I said. These people must all belong to the same country club. So here I was, awarded the biggest Sign of all time. I took the advice of my own attorneys and copped out with a plea of nolo contendere. This was against all my principals, rules of honesty and fairness that I had learned and tried to live by.

The judge looked down and said, "Sorry, Mr. MacDonald. I will schedule a hearing for sentencing. You will be notified of the time."

I heard the gavel.

"What does he mean by sentencing? You told me they would let me loose."

"Don't worry about it, Mac. It's only a formality," Greg said patting me on the back. Glenna was grinning like a Cheshire cat.

Mary was mad. Not a word was spoken on the long flight to Denver. Greg had polled the jury and found they would have voted one hundred percent for acquittal. I felt like I had been let out of a dark box. What I did not know was Glenna had tied a leash to my backside. I would not stray far.

"Glad to have you back, we are close to starting operations," Jim said.

"Where's Richard?" Mary asked.

"I don't know. He got in his camper and drove off the other day. Hasn't returned."

"What happened?" Mary asked.

"We had a little disagreement."

"That's a shame. Just as we are on the edge of getting back in business."

"What's your schedule now, Mac? When do you have to go back?"

"They'll call me when they set the date. Only take three days, then everything will be cleared. It's only a formality."

"The final decision on Mortimer is coming up. I'll be going to Tucson. We should have him back in the family," Mary said.

"Things are in order. The only thing is we are about out of cash. Robert and I pretty much depleted the accounts while you were gone," Jim reported.

"We'll be okay. Spielman will put in more cash as soon as we process some concentrates," Mary said.

"The crushers, conveyers, feeders, grinders and dust collecting system are all ready. All I have left is getting the piping in place on the acid tanks. Won't be long until we start up. By the time you get back, we will be pouring some ingots."

"Good deal, Jim. A lot of work has been done in record time. We're overdue for a change in luck."

There was just enough cash left for Mary's trip to Arizona and a round trip ticket to Atlanta. I had a feeling of euphoria. Everything would work now. By tomorrow, I would be clear of the Glenna problem, Mary would get the plane, and we had a contract with the company in Fairplay. I ordered a scotch, leaned the cramped plane seat back and slept my way through the airways to Atlanta.

I found Greg's boss, John Black, in the courthouse. "Hi. What are you doing here? Where's Greg?"

"Greg has the flu. He's sick. I'll take over the procedure," John said.

"This won't take long will it? I'd like to get a flight back this afternoon."

"This will be a snap, Mac, no sweat. We'll be out of here within the hour."

The judge was staring at the document in his hand. "Humm, case number 3562. What was this, Glenna?"

I had thought this trial against MacDonald was an important event. "How come I've been reduced to nothing but a number, John?" I whispered.

"Shhh," said John.

"Your honor, this is the case relating to the Goldstein problem, the gold selling scheme," Glenna said.

"Oh, yes. I think I remember that. They were sentenced, weren't they."

"Yes, your honor. Harold received seven years and his brother five. They are serving at Terminal Island at this time."

"Oh yes. Well, if they are serving time, then you should, too," he said peering over his glasses at me."

"But, your honor!" I could not think.

"What do your recommend, Glenna?"

"Three years and five years probation, your honor."

"Well, let's see…" He shifted through some papers. "No, I believe one year, but the probation time will be five. Mr. MacDonald you are remanded to custody until a suitable facility is arranged."

My mind deserted me. When I was able to see again, I found

myself in a holding cell. "That fucker, John Black, hadn't said a word," I said aloud.

"What did you say?" a voice asked. There was someone else in here.

"Sorry, just thinking. Do I know you? You look familiar."

"Yeah, we met here a few months back."

"Sure, I remember now. You were the guy involved with the bullion theft in Chicago."

"That's right. Why are you back?"

"I got screwed by the establishment, I guess. Say, you told me you know about refining, right?"

"Yeah, we are all set up in Colorado now."

"Here's my address. I get out today. Come see me in California when you can. I have an idea that you can help me with that bullion.

"You have it?"

"Sure. It's under the concrete of my driveway. The statute of limitations on that job are running out. I'll be able to do something with it."

After being treated by Glenna and my wonderful government like an animal, it did not matter how wrong it might be to get involved with this. "Sure, I'll help you. Soon as I can, I'll find you and work out a deal."

John Black showed up. He is the guy that should be in this cage. "Okay, Mac, good news," he said. "They're going to let you out of here on your own recognizance. The judge didn't know that you had already spent time."

"The judge didn't seem to know about anything," I groused. "I'm nothing to him but a number on a piece of paper."

"He's awfully busy. All of them are overworked."

"Too bad. I feel sorry for him," I said sarcastically.

"This is wrong, Mac. We'll straighten it out. Here is an address. Meet me tonight and we can discuss it further. I have to get to another client." Poor client I thought.

We met that night at John's. Other than John, I did not know any of the people there. "Have a drink, Mac. I think we can clear up this misunderstanding."

"What do you have going?"

"Well, I play golf with the judge, you know."

I was not sure I was hearing right. "You do?" I said.

"Sure, for ten thousand dollars, I'll whisper in his ear and he'll let you go with no sentence."

I lost it. "Why you no good, cocksucking, miserable son-of-a-bitch. How come you didn't whisper in his ear before?" I hit him in the mouth as hard as I could! The party was over. I did not even get to finish my drink. John Black did not sue me for assault. He knew he was wrong.

I drove across the bridge at Bonanza. Something was wrong. The mill building seemed to have a black aura around it. "Mary, what has happened?" She had a coffee cup in her hands and was just staring at something far away.

"Mac, I don't think I can take much more. Something has to change."

I had never seen her like this. "Where's the jug? Let's have a drink," I said. I found a quart of Jack Daniels and poured. "Now tell me what can be so bad."

"They took the machinery, the tools, everything."

"What do you mean, they took the machinery. Who?"

"Jim and Barney. When I got back from Tucson, I caught them with trucks and trailers loading everything. They didn't expect I would get back so soon."

"How in hell could they do that? Did you try to stop them?"

"I went up to Gayle's and called the Sheriff. Then parked the truck on the bridge so they couldn't get out." She took a drink of whiskey.

"Where was Jim?"

"I didn't see him, but his son, Bob, was helping. One of Barney's goons came down with a cherry picker and told me he would drop my truck in the river if I didn't move it. If I had had my gun, I would have shot him. He was awfully nasty."

"What happened to the Sheriff?"

"He and the deputy finally showed up but they were on Barney's side. Barney and Jim had given them the contracts and explained they were just moving the stuff to Fairplay. They knew we wouldn't go along with the idea."

"But how can this be a legal move?"

"The law in Colorado says one partner can make business decisions without the knowledge of any of the other partners."

"That's neat. Jim sold out to Barney. Why, I wonder?"

"Greed, Mac. Pure greed. We owned twenty five percent of the contract."

"But I've known Jim for over twenty years."

"When money is involved, friendship ceases."

"Yeah, my dad put it this way, 'Where a dollar is involved, hen shit freezes'. I would have thought the Sheriff's department would have looked closer at what was happening." I could have never contemplated this kind of betrayal. The shock of what Jim Scott had done to us was so outrageous that I would repress it for some time.

"Mac, you've forgotten that we are arch criminals, enemies of society."

"Yeah, that's right. Us bad guys are going to rule the world. What's left in the building? Do I dare go look?"

"I wouldn't. At least not today. You'll only get sick."

Our jug of whiskey was still half full. "Have some more. Tell me about the trial in Tucson."

I poured more Jack Daniels in her glass.

"You'll like this, too. The judge awarded Morales thirty thousand dollars. Dick gets to pay that. The plane was given back to me."

"Well, that's a plus. At least we have something to show for..."

"Hold on. I'm not finished," she took another swallow. "Our attorney took Mort in payment for what we owed him."

"Goddamn! Morales picks up thirty thousand cash, the attorney gets the plane and we get the shaft. Next time we'll get a bigger bottle of 'ole Jack'."

"If we keep up the pace, we'll need a barrel," Mary said. "I learned also that Morales ripped off Burkett and all of his investors. They say he left town with over a million."

"He's a pillar of society, ain't he?"

"So what's with Atlanta, are you free?"

"You're not going to believe..." When I got through with the story, we had finished the whole jug.

"Mac, you've got to make a quick decision on this." I was on the telephone with Greg my Atlanta attorney.

"What's happening? What options do I have?"

"I've got these people to agree to a ninety day sentence. I don't want to push much further. They could press for a large fine."

"What about John Black, my buddy?"

"Forget John, I quit the firm. I'm working on your case on my own nickel. If you agree to ninety days you won't have to go into a federal prison."

"Why not?"

"It takes longer than that to process the paper work. You will go to the county system under a federal program."

"Where would I go? Here in Colorado, where?"

"You have three choices, Boulder, Colorado Springs, or Pueblo."

"Are these maximum security jails?" I was an expert now. I could ask important questions like that. "Do I get to play golf like Goldstein?"

"No, not maximum security. You will have it easier than the county prisoners. They treat federal prisoners better. That's the law. Probably be no golf though."

"What do you think, Greg?"

"Take it, Mac. Get it over with and get on with your life. After all it's only ninety days." Greg had not been there and done that. Life in a cell for one day let alone ninety was an eternity.

"Okay, set it up, Greg. We don't have much choice."

Mary delivered me to Denver. I was to present myself to the Marshall's office. "How much money do you have, Mac?"

"I have a two dollar bill, a silver certificate."

"One advantage you have is free room and board."

"Can you manage for three months?" I asked.

"Don't worry, I'll manage." I learned later she had twenty dollars in cash. She spent half on gasoline to get back to Bonanza.

Glenna got to me again. I was assigned to the maximum security section. All my new friends were in for murder, major dope running or worse. This was definitely the place not to sleep at night. It took almost five minutes before everyone knew I was special and they did not like 'federal prisoners'.

"They told me this was a special federal program. They allow me an extra hot dog once a week."

"That's pretty special." Mary was peering through a small window. The jailhouse was open to visitors an hour and a half per week. Visitors were allowed twenty minutes to peer into the little window at their beloved inmates. "I won't be able to come very often. The price of gas has gone up again."

"That's okay. I don't want to see anyone anyway."

"Don't get down on yourself, Mac. We'll get things started again."

"What for? Why try? Some asshole will try to take it away."

"Do you have everything you need?"

"I am allowed a package of Top tobacco every week. I've been doing research on how to roll a cigarette."

"Do you have books?"

"Somewhere. The person that runs the library is on vacation. Can't get any for two weeks."

"Do you get the newspaper?"

"No, the only news is on TV. But most of the time these guys are watching cartoons."

"Gold went up to two hundred seventy. Silver is seven dollars eighty cents."

"You know what, Mary, I don't care."

"Get some rest, Mac. You look awful tired."

"Yeah right, fat chance in here."

"Time's up, gotta go."

"Bye."

It's the shits to be a monkey in a cage.

So how to get the days to go by? Why not write a letter to the judge? Maybe they will let me have a pencil. "Say deputy," I called through the bars, "could I have a pencil?"

"Maybe, I'll check."

Eventually he came and handed me a pencil. It was about two inches long. "What's with a pencil this short?"

"Any longer, you might stab somebody with it."

"Yeah, that's really what I wanted it for."

Then this guy asked, "What do you want it for?"

I could not answer. The question was a tough one. "How do I sharpen it?"

"That's your problem," he said walking away. Well you have your tools, MacDonald. Where do I get paper, I mused. The nice fellow next to me was about to throw a letter in the trash. I got it before he crinkled it up. I crossed out the illegible scribbling and started on the backside. 'Dear Judge.' This would take some thought. Without an eraser, I could not make too many mistakes.

> I am writing this to you to try to get my mind clear as to what happened while I was under the jurisdiction of your wonderful system of justice. The thing I would like you to understand is we were doing legitimate business when we were stopped by an allegation. I had spent seven to eight years developing our mines. Most of the money to do this was our own or from loans from friends and relatives.
>
> My business was closed because of association with people that were obviously in the wrong, but we didn't know that when we started. The worst part is, we were toppled by a government official who didn't seem to care who or what was affected by the exertion of her influence. She apparently went after us to further her own fortunes in life.
>
> Another thing I don't understand is how I could be indicted by a Grand Jury that made its decision based only on information that was one-sided. Why was this secret? Why can't little guys like us have our own say before they issue arrest warrants and assume that someone like me is only beneficial for her. We had no chance to

"How are you doing, Mac?" Mary had come for a visit.

"Not good. I wrote a letter to Atlanta. You take it and mail it on refute any of the allegations. Had we a chance, the government would have saved a lot of money.

For your information, Alphonse Morales was the real control of the whole operation. Still to this day, I don't believe they started the project as a fraud. They intended to make some money. They were smart enough to see the opportunity in the gold operations.

I think that when the SEC started pushing them, the decision was then made to take some money and run. My mistake was that I did not do the same thing. If I had, we would be very rich today. But we were not brought up that way. I was taught to finish a project once it was started, not to quit, run and hide.

Your honor, the only reason I accepted the plea of nolo was because we had no money to defend our position. Had the government given us just ten percent of what Glenna spent everyone would have won. I am very disappointed in what I learned of the way our government is run. The principles of our original forefathers have been trashed. If we have people with the attitude of Glenna Stone, then the general public, the little guy with no money, has no chance.

The main thing I can see is Glenna did not only destroy my group, and me, she knocked over several hundred investors. These people will never know the real reason behind their loss. In my opinion, Glenna, with your help, has saved the world from the 'poison beans'.

Sincerely,
Eugene MacDonald

the outside. I don't trust this outfit. They could throw it in the garbage."

"I don't like to see you like this."

"What's happening on the outside?"

"I got a letter from Richard. Mrs. Brown's attorney in California went after him. He may go to jail, too."

"No shit!"

"No shit. He also said the reason he left Bonanza was he discovered Jim and his son, Bob, were on Barney's payroll."

"While we were paying them?"

"Yes, I gave them money and all the time they were getting money from the other side."

"Fuck! We trusted Jim."

"Yeah, some partner. My time's up, better go." Mary disappeared from the hole in the wall.

"Damn, Mac! You weren't supposed to be put in there with those guys,"

"Who are you?" I asked suspiciously.

"My name is Smitty. I'm in charge of placement. I've been on vacation for a month. I just discovered you were over here. I can move you if you'd like."

"Can't you just get me out during the day? These guys aren't bothering me and I don't mess with them."

"How about working in the library? You could at least walk around."

"What does it pay?"

"All the books you can eat." I was going to like this guy.

The library saved my sanity. I was out of my cell a good part of every day. I could walk around. I could read a newspaper everyday and the choice of books was good. Some compassionate citizens who did not understand that most people in these places could not read had donated them.

In his office, Smitty was working a crossword. "Say, look at this article. This was one of our mining projects," I said taking the Gazette to his desk.

Smitty read it with interest, "You're kidding? You were working there?"

"We had a contract with the owner."

The article read, "Nome Alaska. A large gold discovery was reported last week. The old Bucket Line Dredge was activated this season on the Solomon River. In just sixty days of operation a total of four million dollars in course gold has been recovered and it is estimated there is much more contained in the black sand concentrate."

"Luck of the Irish, Smitty. I'm locked up here while these guys are running what could have been our project."

"Unbelievable. I am going to look into your case. This is interesting."

Mary generally came down from the hills every two weeks to say hello. I really did not want to talk to anybody and this day was no exception.

"Well, Mac, more news." She really looked tired from the long trip.

"What?"

"Jim Scott is in the hospital. Acid tanks blew up while he was working on them. Actually he was blown through the roof of Barney's new building."

"Jesus. He should have been killed."

"He wasn't. Close for a while though, but he'll pull through."

"What goes around, comes around, doesn't it?"

"Sure does. Nature has her ways."

A couple of days later I was sweeping the floor near Smitty's office and he waved me in.

"Mac, you've been around here and obviously have made some observations. What do you think of El Paso County's state of the art prison system?"

"You want my opinion?"

"Yeah. I'm writing a book on the gains that have been made in the way the criminal is treated."

"The whole thing is nuts. I think all you are doing is providing free room and board to a bunch of characters that have no intention of changing their lifestyle."

"How is that?"

"I've been in three institutions. Each one had a group that knew each other. When they get together, they learn from the other guy what he did wrong to get caught. It's like training school. When they get out, they know more about bank robbing, car stealing or dope running than they did before."

"Interesting."

"The taxpayer is taking it in the shorts. They are actually financing a private school system to turn out a smarter criminal."

"What else, Mac?"

"The way we're treated and looked down upon by the prison officials helps you to start thinking like a criminal. If you weren't one when you come into the system, you will be one when you get out. Part of the problem, in my opinion, is the people you have operating these jails are of the same intelligence level or less than the prisoners. Just because a lot of the inmates can't read and write doesn't mean they are dumb." Over the long weeks, we had many discussions about this subject. The consensus seemed to be that the system did not work.

"Hello, anything interesting to read?" I was in the library studying a book about the reversing of the poles of the Earth.

"Sure, can you read?" He was a good-looking clean-cut guy.

"Yeah, a little," he laughed.

"What are you in for?"

"Aw, I got caught with a shipment of marijuana out of Mexico. They gave me twenty years."

"Twenty. What are you doing here in this lowly place?"

"I was in Canon City at the state pen. They sent me here for protection. I may get a reduction in my sentence."

"What happened?"

"I learned the guards were making dope buys on the outside and smuggling to the inmates. I couldn't see why I should be in while the same people that put me there were dealing."

"So you ratted on them."

"I made a deal. Helped the narcs run them down in trade for my time. I'm up here for my own protection. They would have killed me in that bastille."

"How long were you running the border?"

"About three years. It was easy. Made a lot of money."

"I know the border. How did you last so long?"

"We had better radios and checks than the fed. We had their

frequencies and knew where they were all the time."

"How did you bring it out?"

"Mostly by plane. We had a deal working where we could get our own numbers and colors changed. Can't tell you about that though."

"You don't have to. I think I know what you're talking about. One thing though, did you fly a Queen Air?"

"Why, yes, how did you know that?"

Amazing, I thought, the world gets smaller every day. That was his plane we had seen in Culiacan.

On my last day I was kept waiting to be released until close to five o'clock. They had neglected to give me back my two dollar bill, but I decided not to argue too loud about it. "What happened in there?" Mary asked. "I've been waiting since nine o'clock this morning."

"I told you I was a special prisoner!"

"Uh huh."

"I was special because the feds subsidized them to keep me. I was worth three hundred dollars a day. Had they let me go earlier today, they would have lost a day's pay."

"So at that rate, the cost to keep you for ninety days was twenty seven thousand dollars."

"Yeah, I wonder what the total cost of this has been. So what else has happened?"

"I was going to save this and not tell you 'til later. Seems like Burkett has been trying to locate the grinder."

"What for?"

"I talked with Clay. He says that Roy has been making a noise around Tucson that you stole his machine."

"That old bastard ought to have his balls cut off! Oh well, just one more piece of bullshit."

"Yeah, just one more piece, but I wonder when this will stop," Mary said.

"Well, it ain't stopped yet. There is still this probation thing."

"What are you thinking? I sold everything that was available. We have no more money."

"I'm going to find the Coors guy, Bernie Meinerz. He might help. Said he would."

"I didn't tell you about Jim."

"My good buddy, I've tried to repress that. How's he doing?"

"They screwed him. When he got out of the hospital, Spielman fired him. He and Bob went back to California."

"So, one more exercise in futility," I said.

I had forgotten how peaceful the mountains were. "It would be easy just to sit here and look at the trees forever," I said pouring some real coffee.

"Yeah, but we can't eat the trees. We had better get something going. Winter's not far off," Mary reminded me.

"Okay what do we have to work with?

"I was too upset to take inventory. I plugged up some holes in the wall and locked the door."

Jim had left my tools, some acid and beakers in the lab. We had a drill press and a welding machine. It had taken Mary several days to organize the few remaining pieces and parts and clean up the mess the thieves had left.

"We'll have to go across the creek to get more firewood," Mary said. "There are a lot of aspen that are down over there."

"Who's that coming up the road? Do you recognize that car?"

"No. That little thing doesn't belong up here. Wonder how it got up the hill."

"Hello. Can we help you?"

"I hope so. I'm trying to locate Eugene MacDonald."

"Are you from the Sheriff's department?"

"No."

"Okay, then I'm MacDonald."

"My name is Walt. I'm from Denver. I'm to be your probation officer."

"Why are you out here?"

"My job. I'll be checking on you once a month."

"For what?"

"To make sure you stay out of trouble. I want to interview you. I have a report to make."

"Okay," I sighed, more bullshit.

"Come on in the house. I'll make some coffee," Mary said.

"Oh no! I can't do that. Regulations say I can't have any personal relationship with you."

"That sounds like what I'm used to hearing, more government crap." I replied.

"I have a list of things you're not to be involved with. We'll go over them together."

"Give it to me. Let's see it."

"Number one item says I can't have a gun?"

"That's right, no guns."

"That's bullshit. We aren't living out here in the mountains with no weapons."

"If you keep a gun, I'll report it as a violation."

"Report it then. We keep our guns. Number two says I'm not allowed any explosives."

"That's right. You'd be a threat to society."

"I'm in the mining business. I need dynamite. Number three says no chemicals. Number four, I can't associate with anyone until you clear them. Tell you what, you take this list and stick it up your ass."

"Mr. MacDonald, you can't talk to me like that. I can put you back in custody."

"You go ahead and do that. You people have destroyed us already. Now you're telling me I can't do what I know how to do. You personally are going to control my existence for the next five years?"

"Mac, calm down. He has a job to do," Mary admonished.

"Yeah, you're right, but so do we. How much do you get paid to come out here, Walt?"

"I get twenty five dollars a visit."

"You're kidding me. It takes more than that for gasoline."

"Yeah. I'm not going to make much from your case."

"Listen, you're dealing with bad guys, right?"

"Yes. Some of them are dangerous."

"I'm not one of them. Sit down there and I'll explain what really happened to us. Then you make up your mind."

I told him the story. He left looking thoughtful.

Bernie Mienerz was true to his word. He loaned me three thousand dollars and gave me an electric motor and a conveyor. Some how, some way, we were going to build another grinding machine.

Mary kept busy cutting wood and kept the fire going while I built the machine. This was a good thing, having something to do. I could cut plate steel with the torch and drill holes. I had a big sledgehammer to hit things with, and if it broke, I could weld it back together. The mental concentration and the physical exercise allowed my nervous system to readjust.

My good friend, the parole officer, showed up on schedule, exactly thirty days and almost to the minute.

"Hey Walt, good to see you. I need some help here. Grab on to that motor and set it on the bench."

"I can't do that. I'll get my hands dirty."

"Use some gloves then. If you're going to come around and bother me, I'll put you to work."

"I wish I had the time to help you. I would like to learn how to do some of the things you do."

"How is my sweetheart, Glenna Stone? Have you checked with her?"

"It's rather interesting, Glenna was fired from her position. You were her one and only case."

"No shit! I really feel for her."

"She's teaching now at Emory."

"Those that can't, teach," I replied.

"I've been looking into this. Something is not right. You and Mary aren't criminals."

"Tell the world that, Walt. We can't get a loan at the bank and I can't get a job because of my record."

"I'm going to do more research. I think I'll write to the judge who handled this and learn what he thinks. Mac, I think I can trust you. I have an idea that will save me a lot of time."

"Oh?"

"I've made out these monthly reports and dated them in advance. If you will sign them, I won't have to drive all the way down here."

This guy should join with Morales, I thought.

Mary cut wood and I cut steel. Her problem was keeping us warm without using up our inventory of trees. My problem was not making any mistakes. There was just enough steel to get the machine together.

"I found out how Mary Taylor makes ends meet," Mary said. "She has a real estate license."

"That old girl is something. At eighty-four she is outdoing me with energy. But what can she sell around here?"

"She sells the cabins that the summer time people own to the skiers in the winter. According to our neighbor, Dutch, she's sold the town at least four times," Mary laughed.

"She must have been a swinger in her younger days."

"Yeah, she was. Dutch said she would find a miner with money and marry him. Soon as the money was gone the miner would disappear. Happened several times."

"Interesting coincidence, huh!"

"One of her husbands committed suicide. The police report said he blew himself up with a stick of dynamite."

"That would be hard to do. Just hold onto some powder and watch the fuse burn."

"Seems it wasn't quite like that. Dutch said Mary had been seen coming out of the house that night. She put the dynamite under his pillow while he was asleep."

"Jesus! You haven't found where I hid our dynamite have you?"

"Not yet. I'm still looking though," she said as she threw another log in the stove.

"That reminds me. I've worked out a plan for Morales."

"You should not dwell on those guys, Mac."

"I know but this idea would be helpful to him I believe. Make him think."

"Okay, what is it?"

"Remember that building we found in Sonora where the airplane was parked?"

"You found the fuel there for Mortimer, I remember."

"There was a bunch of old rusty barbed wire laying around."

"So, what's the plan?"

"I'll kidnap Alphonse and take him down there. This is a good spot because it's so hot. I'll lay him on a block of ice and tie the barb wire on the overhead beam and around his balls."

"Are you planning to stay and watch the ice melt?"

"No. I don't think so. It's better if he is left alone with his thoughts."

I finally got the machine together. I slipped the V-belts onto the grooves of the sheaves and adjusted them to the proper tension. "There she is, Mary, a new creation."

"It's beautiful. If I had some paint, I could finish it for you."

I took hold of the belts and gently turned the rotor system. My fingers were pulled into the main drive pulley. "Ow! Damn it!' I screeched.

"What's wrong. What did you do?"

"I broke my fucking finger."

"Damn, Mac, that's bad. Why didn't you break one of the others?" Mary grinned.

"Man, that hurts," I complained sitting down on a carbide can.

"Shall I call 911?"

"Go ahead. We can pay for the ambulance with the million dollars I hid in Switzerland."

"Don't you wish. Let me see that finger." Mary pulled on the crooked finger and we heard a loud pop. "Well, at least it isn't crushed." She found an old teaspoon, broke off the handle and used it as a splint, wrapping it tight with electrical tape. "There you are, good as new."

"My goodness, it's cold out there." It was Walt. "Do you have any coffee?"

"Aw come on, Walt, you said that drinking with the enemy was against the rules."

"It's okay. You're not on the most wanted list anymore," he laughed.

"What's up, Walt?"

You know what?"

"What?"

"The judge agreed with me. He turned you loose. He said you weren't guilty of anything!"

"You're kidding. After all this time..."

"What's more, the whole proceedings have been expunged from the records. None of this in relation to you ever happened."

"I have a bit of a recollection about it, Walt."

"Well. Forget it. You're a free man!"

Mary looked over at me and cried big tears for both of us. The quakies rustled in the breeze. They could not care less.

⌘ ⌘ ⌘

25 Years Later - Medford, Oregon

I woke about 4:30 a.m. A crescent moon and a sky full of stars greeted me. It was a dry moon and the morning was quite cool. A good day for driving. Ole' Herkimer, the Cadillac I had purchased from Richard in 1974, had a full tank of gas and was fit for another journey.

I hit I-5 North. Since it was Saturday, I found light traffic volume and few trucks. This was going to be an easy run to Vancouver, Canada, I thought. I had driven the route many times, and I knew almost every rock and tree by name. Generally, it took about nine to ten hours, depending on the traffic in Portland and Seattle.

I was thinking about writing another book and some of the information was in my computer, at Christina's place, in Vancouver. I had met Christina ten or eleven years ago and she allowed me to stay at her townhouse when I was in town.

From Southern Oregon, the Canadian border is about 600 miles. It was a pretty drive through the trees and once I got past Everett I could see the tulip fields in bloom this time of year.

Herkimer was running well, but the big Cadillac did need fuel from time to time. I stopped in Woodland, Washington, and filled up with $33.00 of gasoline. Now it was only about four hours to the border crossing at Blaine.

I arrived at the Blaine crossing about 2:30 p.m. and was feeling good, but tired, of course. The odometer read 570 miles, but it had been a fairly, easy journey. There were only a few cars waiting to cross into Canada, and most of them were from Washington State. With the advent of the twin tower debacle in New York last September, the higher security was apparently discouraging a lot of unnecessary travel. Not a lot of Canadians were coming to the States because of the increased security. As usual I gave the girl at the checkpoint my passport and told her I was going to visit a friend and if the weather was nice maybe practice some golf swings.

"When did you visit last?" she asked.

"I was here last March, I think."

"Did they do a check on your passport at that time?" she inquired.

"No, they did not. It was checked by the Chinese Consul last summer when I went to Beijing." I thought that would stop her questions. Obviously, the Chinese would have turned me down if there were any problems with my identification.

She handed me my passport with a card stuffed inside and said, "Please park over there and take this inside to the immigration desk."

I did not like the delay but I had done this several times before without any problems and besides who would bother with a grey-haired old man carrying around 68 years on stooped shoulders. Inside, I approached a desk where a large sign read Immigration Desk. There were maybe eight people standing in line, some with hip huggers and tee shirts. Look like the hippies of yesteryear, I thought. Other than one official looking kid behind the counter typing away on a computer, nothing seemed to be happening. Standing in the line waiting, I began to feel the fatigue of 570 miles driving creep up. I told myself, hang on, I will be out of here in a few minutes. After about twenty minutes, all the hippies had passed through, and it was my turn.

A young girl took my passport looked at the card inside and asked, "Have you ever had any trouble with the law?"

"No, I have not," I answered.

She disappeared into a back room for two or three minutes. Upon her return, she asked, "What is your social security number?" I gave her the number and she said, "Are you sure you never had any problems with the law?"

"Years ago, in the 70's, there was a problem, but it was taken care of," I answered.

"Well my information shows you were charged with mail fraud in 1977," she said.

"Yes that is true, but it was all taken care of," I answered.

"Under Canadian Law, if you have any criminal record from the past you are not allowed to enter the Country," she sneered.

A flash crossed in front of my eyes and I saw Glenna Stone, all over again. "Listen Sweetheart, I have been coming into Canada for twenty years. I spend a lot of money here. I have been trying to do business with you people. Why are you stopping me now?"

"You can't call me sweetheart. Obviously nobody has checked on you and I have found a record. You are not welcome here and will leave immediately. We don't need your kind in Canada," she said handing me my passport and another sheet of paper.

"What if I don't go back?" I asked.

"The RCMP will be on your ass immediately," she snarled. "Give me your passport."

"No, I won't give you my passport. It's mine." I had learned years before never to give anyone your passport.

"You come back here, I am not finished with you yet," she yelled across the room, as I walked out to the car to get my glasses so I could read what she had given me. Leaving the car door open, I sat down with my feet outside of the car and put on my glasses and started to light a cigarette. The first words I saw said, "ENTRY DENIED," and then, "EXPELLED." Now what the shit is this, I thought, starting to read the rest.

"Get out of the car, smart ass. Do you want to go to jail?" said a voice.

I looked up and saw a young guy in uniform and some people standing behind him.

"You know, that might be a good idea," I answered.

"Okay, smart mouth, you are under arrest." He grabbed me by the left arm and jerked me out of the car. He tore the cigarette from my mouth and said, "Drop what you have in your hand." It was my lighter. "Put your hands behind your back. Spread your legs." Then he put handcuffs on me. Then he searched in my pockets and emptied everything on the roof of the car.

"You can't carry anything sharp," he said when he found my pocketknife. He threw it into the backseat of the car.

"Do you want to go back, or do you want to go to jail?" he yelled.

"I'll go back. You are hurting my arm." I had not said a word until then.

By this time, the girl was there with another female officer. "Give me your passport. You will leave over there to the US side. I will hand it back when you pass by," she said.

I was rather embarrassed, standing there in the parking lot in front of the incoming tourists with handcuffs on. But I was more embarrassed at the way Canada's finest had purported themselves. In a just a few short minutes, they had destroyed my twenty years of good feelings about the Canadian way. If they had not found a problem with me for twenty years until today, they had pretty, piss-poor security established. I wondered if they ever find a really bad guy.

So back to the States it was. The nice lady handed me my passport and I forgot to say thank you or have a nice day or anything like that. I eased my way into the line of cars waiting to get out of Canada and approached the gate. The guy says, "Hi, where do you live."

I told him and then said, "They threw me out. Said I was unfit for their country."

"Well, hell! Did they give you a paper?"

I gave him the paper and he said, "You will have to take this and go over to that building. See Customs, they will take care of the necessary work. Hope to see you someday."

By now, it was a little after 3:30. I went into the building and found the customs counter. After a few minutes, a gentleman showed up and said, "Okay, lets see what problem you have." I handed him my passport and he disappeared for a couple of minutes. He came back and asked me where my car was parked. We could see the car from the window. "Okay, just turn around to the left and head on out."

"Thanks. What am I supposed to do now? They did not tell me a thing about procedure."

"You will have to go to the Canadian Consul in Seattle and apply for re-entry," he answered.

Outside I saw a telephone booth. I thought I had better let Christina know I would be late for dinner. I called collect and asked her if she could come across the border and bring me a clean shirt. I would call her back and tell her where I would be staying. I drove back into Blaine and found the International Motel. It had a lounge. The room was $51.00. Too much, I thought. Nevertheless, shock was beginning to set in and I needed to sit down a moment and regroup. From the room, I again called Christina collect with the name of the motel and for her to look for me in the bar.

After she arrived, she told me of the difficulty in answering all the questions at the crossing. We discussed what to do next and how I could collect my computer and other things that had accumulated over the years. We decided I should go to Seattle and see if they could handle the problem there. After all, it was quite simple, probably a case of new people on duty who did not know the law. I showed Christina how to get back to the Peace Arch Crossing and went to Seattle Sunday morning.

The drive to Seattle was terrible. High winds and heavy rain made it treacherous. It seemed to me that everyone was driving faster as the weather worsened.

I found a place to stay the night, but sleep was not on the agenda. The next morning, not knowing where to find the Canadian Consul, I took a cab, and son of a gun, the driver could barely speak English. I arrived before they opened the doors and was tenth in line behind Chinese, Koreans, Rag-heads, and Push-buttons, all getting entry visas for the great world of Canada. I got to the window finally and another young girl says, "There is nothing we can do now. Here is a form to fill out." Thumbing through the document, I discovered ten pages of directions on how to file for re-entry and, of course, how much to pay.

"I want to talk with somebody about this. This whole episode does not seem right to me," I said.

"There is a new law coming out next month dealing with this type of expulsion, but we are not trained to answer questions about it. Give me your passport and have a seat. I will get someone to talk with you," she said.

After about ten minutes they called me to a window, told me it would cost $1000 to file the rehabilitation paper work, and that it would be six months before the processing would start. The best thing about this was the money was not refundable, and they did not have to allow me rehabilitation status. Again this came from a young female. I began to wonder if women were running the whole world.

"What about my computer and things I have in Canada?" I asked.

"Just have someone bring them to you." She said, shrugging her shoulders.

So that was it—out the door and into the cold. If it were not for high profile terrorists like me running around and trying to destroy the world, what would these people have to do? It was coming clear in my head that Glenna Stone had had the last laugh. So much for adventures in Canada. Maybe I should try China.

The End

EPILOGUE

Mac - *went on to hone his skills with prospecting, mining, metallugry, and different recovery processes. Choosing honest partners, he continued until the advent of environmental requirements and impact studies that cost more than developing a mine pushed him out of the mining business. He acquired the rights to the garbage machine from Roy Burkett and developed processing methods for waste materials.*

Mary - *fifteen years later, was diagnosed with cancer of the spine. She died in a nursing home after being under intensive medical care for five years.*

Mortimer - *Mary's attorney retrieved the plane but took is as payment for his fees.*

Richard - *was put in jail by the attorneys for Mrs. Brown. The chess set was never located. He later married and is now building houses in North Carolina.*

Dick - *paid $30,000 to Morales for stealing Mortimer. The court ruled he should have returned the airplane to Morales, therefore, he stole it from Morales.*

Clay - *went back to Mexico to work on the mine he and Mac had developed. Later, he and Mac worked together on a gold mine in Oregon. Clay died several years later in an accident driving a dump truck up the narrow trail to the mine at the ranch in Mexico.*

Harvey - *took a job as a bush pilot in South America. No one ever heard from him after that.*

Wilbur Wright - *purchased Mary Martin's property in Tucson when Mac and she moved to Colorado.*

Jim *- after several weeks of recuperation in a hospital intensive care unit because of the acid tank explosion, went home to his wife and retired. He later told Mary that his betrayal of Mac's friendship was the worst mistake of his life.*

Glenna *- the Federal Assistant District Attorney, was asked to step down from her position. She went on to teach law at Emory Law School in Atlanta.*

Alphonse Morales *- promoted the grinding machine and made a million bucks by ripping off investors in Tucson.*

Harold Goldstein *- was released from prison for good behavior after two years. He went on with another scheme involving low cost gasoline and was re-incarcerated soon after.*

Roger *-disappeared when the indictment was made against Mac and him. He never surfaced again.*

Bob Slagle *- was one of the witnesses that helped the grand jury make its decision to indict Mac. Later Slagle sent a letter to Mac's father saying if he was not repaid for his loss of the project in Nome, he would arrange to set fire to the brush near his house. Mac gave the letter to the FBI who contacted Slagle and suggested if he continued harassment of Mac he would find himself in prison. There was no further contact with Slagle.*

Walt *- the Probation Officer became a good friend to Mac and Mary. He quit his job shortly after finding out about the government's role in prosecuting Mac.*

Mark Piller *- the Doctor who invested in the mill at Bonanza and loaned Mac the $10,000 to hire an attorney remained good friends with Mac and Mary. In later years he developed Alzheimer's Disease and had to give up his practice.*

Enrique Monesterio *- never got Roy to furnish the information or drawings on the machine that he had purchased for $10,000. He asked for his* *money back from Roy. Roy denied having ever talked to him. Enrique won $2 million in the Mexico lottery three years after saving Mortimer, Mac and Harvey.*

Bishop Jim Stokes - *later helped finance a mine in Oregon for Mac. The property in Saugas for which Mac had a contract to purchase for one hundred thousand dollars later sold for two million dollars.*

Griffin Ranch (the Bonanza Property) - *The owners finally put the property up for auction when Mac and Mary could no longer make the payments. No one bid because the EPA had placed a condition on the property that the owners were reponsible for cleaning up the acid water in the creek that was flowing from the mines above the property.*

The Sierra Pinta - *was sold by Montgomery to another American who installed a cyanide plant to do the extraction of metal using the pumps, generators and tools Mac had taken there. Montgomery soon after died of a stroke.*

The Nome project - *was eventually stalled by the environmentalists.*

The Gold Lake Mine - *was opened by a Canadian company. It was reported they made some money from the gold found there.*

ABOUT THE AUTHOR

Donald E. McCoy graduated from the University of California, Santa Barbara in 1959 with a degree in Industrial Management. He and his brother took a trip to Baja, California, in 1956 driving two Renault 4CV's. They set the first speed record to Cabo San Lucas. The record was twenty-one days. In 1958, they drove a 1934 Model C Ford pickup to Central America. They traversed what is today the Pan American Hiway. The fun they reported of talking their way through Castro's forces—as they were training for the attack on Cuba—is a story all its own.

Running out of money Don returned to California and took a job teaching school to earn enough funds to send to Costa Rica so his brother could get home. After a few years of teaching school and building houses, he became interested in the romance of gold mining. The story here is based on one of many adventures he has had. He wrote about this adventure because it covered a wide area of different levels of business in the United States and Mexico.

Don lives in the Pacific Northwest where, if you search about, you can still find a big tree and a running stream. He is still into real adventures as well as ones on his computer.

Other Books Available
From
BLUWATERS PRESS

MASSAGE DURING PREGNANCY, by Bette L. Waters. The first book on the market written about pregnancy especially for the massage profession, is now available in new revised and expanded Third Edition. **Home study for 12 CEU's.** Approved as Category A by NCBTMB and the Florida Board of Massage. It is crammed full of what the massage therapists needs to know in order to safely massage pregnant clients: anatomy and physiology of pregnancy; discomforts of pregnancy and how the therapists can help with relief; massage basics; psychological adjustments of pregnancy; labor support; postpartum period; marketing; research.

"Packed with essential information dealing directly with massage and the pregnant client."

Sue Welfley, LMT, former President Florida State Massage Therapy Association

"...wonderful reading for the mother-to-be as well as for the massage therapist...a must read for everyone involved in the healing arts...the kind of book that makes a difference."

Paul St. John, LMT

HIGH SEA PASSAGES by Billy A. McClellan. McClellan and his contemporary, Harry Crews, grew up in Bacon County Georgia graduating from Bacon County Highschool in the 40's. His true-to-the-bone characters are drawn straight out of the southern culture he went to sea to escape, only to return five years later and write about the southern male as only a southern can.

"A tale of 29 southern 'cowboys' hired to handle 600 horses on their way from Georgia USA to Greece on a U.S. Merchant Marine Victory Ship. The cowboys range from naive 17 year old farm hands on their first trip away from home, to middle aged men with painful pasts. *High Seas Passage* is a remarkably poignant and humorous story of this cowboy crew, with the younger ones gaining confidence in new adventures on ship and in Greek ports, and the older cowboys finding answers to life-crippling issues. From shoveling tons of manure, craning dead horses off the ship in mid ocean, drinking, gambling, arguments, fights, and sexual exploits, the cowboys make the most of their voyage...this book is an important, entertaining snapshot of how a group of fascinatingly diverse people cope with the passages of life."

Robert Ring, *History of Ruby Arizona*

"These characters explain why southern females are called upon to be "steel magnolias".

Bette L. Waters, *Massage During Pregnancy;*
My Daddy Brought Me Up To Be Good

"Somehow the sea is a place for the worst to better themselves. It accepts all, like a big and powerful God."

The Vet, *High Seas Passage*

"This is a sea story with an uncommon pallet of characters. It represents the wanderlust of a young man's heart and the search for relief by a man whose life has left him with a hole in the fabric of his soul...a story that captures a wide medley of varied motivations while sketching a portrait of a slice of post-World War II Americana long since past."

Captain John C. Timmel, Harbor Pilot
President The Victory Ship Incorporated
Tampa, Florida

MY DADDY BROUGHT ME UP TO BE GOOD by Bette L. Waters. Her first book of poetry is the story of one woman's triumphant journey towards self-realization.

"Bette Waters' selections sparkle like clear crystal drops of memory and remembering. I weep and rejoice with her as I feel my own womanhood through her words. They are delicious reflections of girl and woman—mine, hers, and ours.

Elizabeth Rauh Bethel, *Promiseland, A Century of Life in a Negro Community; AIDS: Readings on a Global Crisis; The Roots of African American Identity.*

"As a child of southern culture we all love our secrets. It is a culture that does not encourage vocalizing. Bette Waters has broken an old southern rule—she had the courage to TELL."

Dr. Janet A. Sipple, Professor,
Moravian College, Bethlehem, PA.

ORDER FORM

BLUWATERS PRESS
P O BOX 878
MESILLA, NM 88046

QUICK ORDER: Call 1- 888-541-5381 Toll Free MC/Visa

VISIT WEBSITE: www.zianet.com/bwaters

CONTACT VIA EMAIL bwaters@zianet.com

$19.95 [ ] One Foot Away From A Million
$18.95 [ ] Massage During Pregnancy
$125.00 [ ] Massage During Pregnancy Home Study CEU's
$16.95 [ ] High Sea Passages
$7.95 [ ] My Daddy Brought Me Up To Be Good

Name:____

Address:____

City:______________________**State:**______**Zip:**________________

Telephone:________________**email:**__________________________

Sales Tax: Please add 6.5% for purchases shipped to New Mexico.
Shipping $5 one book. Add $1 for each additional book

Priority Mail USPS or UPS
International: USPS Global Priority or USPS Airmail

Payment: [ ] Check [ ] MO [ ] Credit Card

Card
Number:__________________________**Expiration**______________

Name on
Card________________________________**Amount**___________

Printed in the United States
5694

9 780966 558463